Mathematics, Poetry and Beauty

Mathematics, Poetry and Beauty

Ron Aharoni

Technion, Haifa, Israel

NEW JERSEY • LONDON • SINGAPORE • BEIJING • SHANGHAI • HONG KONG • TAIPEI • CHENNAI

Published by

World Scientific Publishing Co. Pte. Ltd.
5 Toh Tuck Link, Singapore 596224
USA office: 27 Warren Street, Suite 401-402, Hackensack, NJ 07601
UK office: 57 Shelton Street, Covent Garden, London WC2H 9HE

Library of Congress Cataloging-in-Publication Data
Aharoni, Ron.
[Matematikah, shirah ve-yofi. English]
Mathematics, poetry, and beauty / Ron Aharoni, Technion, Israel Institute of Technology, Israel.
pages cm
ISBN 978-9814602938 (hardcover : alk. paper) -- ISBN 978-9814602945 (pbk. : alk. paper)
1. Mathematical recreations. 2. Mathematics--Study and teaching. 3. Poetry in mathematics education.
I. Title.
QA95.A27413 2014
510--dc23
2014026663

British Library Cataloguing-in-Publication Data
A catalogue record for this book is available from the British Library.

Hebrew edition: *Matematika, shira veyofi* (Hakibbutz Hameuchad Publishing, Tel Aviv, 2008)

Translators for the English edition: Merav Aharoni and Edward Levin

Typeset by Stallion Press
Email: enquiries@stallionpress.com

Printed in Singapore by Mainland Press Pte Ltd.

Contents

Introduction: Magic

A mathematician who is not also something of a poet will never be a complete mathematician.

Karl Weierstrass, German mathematician, 1815–1897

Mathematics and Poetry

Poetry is the expression of the imagination. In it diverse things are brought together in harmony instead of being separated through analysis.

Percy Bysshe Shelley, English poet, 1792–1822, *Defence of Poetry*

Mathematics is concerned with understanding the differences between similar things, and what is shared by different things.

James Joseph Sylvester, English mathematician, 1814–1897

The moving power of mathematical invention is not reasoning but imagination.

Augustus de Morgan, English mathematician, 1806–1871

The great German mathematician David Hilbert (1862–1943) noticed that one of his students had started missing his lectures. When he asked for the reason, he was told that the student had left mathematics in favor of poetry. "Ah, yes," said Hilbert, "I always thought he didn't have enough imagination for mathematics."

Hilbert's derision of poets should be taken with a grain of salt. After all, his attitude toward physicists was not much better: he once declared that "physics is too hard for physicists." But he was not the only one who compared mathematicians with poets, in favor of the first. Voltaire, for example, said that "there was more imagination in the head of Archimedes than in that of Homer." Even poets agree. The American poet Edna St. Vincent Millay entitled one of her sonnets "Euclid Alone Has Looked on Beauty Bare."

This poses a riddle. How can the austere and abstract world of mathematics resemble art? What does geometry have in common with music, or arithmetic with poetry? One answer is that both mathematics and poetry search for hidden patterns.

> *This poem is a poem on people;*
> *What they think and what they want*
> *And what they think they want.*
> *Besides this, there aren't many things in the world*
> *That we should care about.*

Nathan Zach, "Intro to a Poem," from *Other Poems*

A poem tells us what we really want. And what poetry does to human emotions and cravings, mathematics does to order in the material world. It tries to find the internal logic of things.

But this cannot be the full answer. Every science, exact or not, looks for rules underlying external appearances. What is it about mathematics that makes it more akin to poetry than any other science? There is another, more prominent common feature that makes us feel that mathematics and poetry are close: beauty. Nothing is as practical as mathematics. Our everyday life, which is so dependent on scientific progress, is influenced by mathematics in very tangible ways. This, however, is not the secret of its attraction, not for professional mathematicians, and certainly not for amateurs. Most engage in it for a completely different reason: its aesthetic value.

In this book, I will try to trace those features and mechanisms that are common to poetry and mathematics, causing them to share the same type of beauty. For this purpose, I will have to touch upon that most elusive of philosophical questions — what is beauty? The unique perspective of comparing two fields may offer a clue. The very fact that poetry and mathematics are so far apart narrows the scope in which the answer should be sought. It is to be found in the intersection of the two domains, which is much smaller than each of them viewed separately. The smaller the overlap between the two fields, the narrower the area for the common denominator to be searched.

As I said, the answer cannot be simple. But there is one word that captures its essence: *magic*. The sense of beauty, in both poetry and mathematics, is the outcome of a sleight of hand aimed at concealing what is really happening. Here, for example, is definite magic — one of the poems entitled "Time and Eternity" by Emily Dickinson:

Adrift! A little boat adrift!
And night is coming down!
Will no one guide a little boat
Unto the nearest town?

So sailors say, on yesterday,
Just as the dusk was brown,
One little boat gave up its strife,
And gurgled down and down.

So angels say, on yesterday,
Just as the dawn was red,
One little boat, o'erspent with gales
Retrimmed its masts — redecked its sails —
And shot — exultant one!

What makes this poem so effective is its sincerity. Through it Dickinson lays herself bare in a way she wouldn't even dream of doing in real life. "Adrift," "gave up its strife," "little," "o'erspent with gales" — all these describe her life as succinctly as possible, but she probably wouldn't say these words aloud even to herself. It is the metaphor that makes it possible for her to express all these with such courage. The message is conveyed to the reader only on a subliminal level. The effect is knowing, without really knowing. The poem is telling us something deep without our being fully aware of its meaning. A poem is a pickpocket who instead of stealing, puts something in our pocket without us noticing.

No less powerful is the uncovering of a hidden truth. "Poetry is always a search for the truth," were Franz Kafka's words to which we shall yet return. The poem tells us that what is on the surface is only one aspect of reality. The inner forces are of greater importance. Beneath the little boat overwhelmed by the tempest there is a very brave vessel; and even if it looks like it is about to sink, it spreads its sails and takes off. Could Dickinson's own life be depicted more beautifully than this?

Emily Dickinson (1830–1886).

So, this book is about the magic of poetry and of mathematics, and how close the two are. It is divided into three parts. Part I is about order. We shall see in it how both fields uncover deep hidden patterns. In Part II we shall study common techniques of the two fields. Finally, in Part III I will try to draw conclusions on the concept of beauty.

But first, I want to give a glimpse of what's to come. I would like to describe one sleight of hand.

Displacement

The mechanism I want to discuss is common not only to poetry and to mathematics. It appears in almost every area of human thought. Its name, "displacement," was coined by Sigmund Freud, who discovered it in dreams. Displacement is the diversion of attention from a central figure to a side one. The main character of the play is shunted to the murky edges of the stage, while the spotlights focus on a less important figure. The main idea is thereby presented incidentally, as if offhandedly. In dreams, so claimed Freud, the aim is to conceal some forbidden content, letting the message slip the attention of our inhibitions. This, according to Freud, is the aim of all dream techniques. In mathematics and in poetry the effect is beauty. It is a magician's ruse, telling his audience "Look at what I am doing with my right hand" while performing the trick with his left.

Here is an example in a poem, "About Myself" by the Israeli poetess Lea Goldberg. It is an *ars poetic* poem ("ars" is the Latin word for "art"), meaning that its topic is the poetry of its author. Goldberg examines the connection between her poetry and her life, and reaches a painful conclusion:

> [...]
> *My images are*
> *Transparent like windows in a church:*
> *Through them*
> *One can see*
> *How the light of the sky shifts*
> *And how my loves*
> *Fall*
> *Like dying birds.*

Lea Goldberg, "About Myself," *Lea Goldberg: Selected Poetry and Drama*, trans. Rachel Tzvia Back

The most transparent stratagem of this poem is the metaphor. In fact, it is a second-order one, a metaphor within a metaphor. The poems are compared to images, and the images, in turn, are likened to church windows. But the heart of the poem is in its last three lines, in which the poetess tells with painful sincerity of the fate of her loves. I live in my poems, she divulges, while in reality, my loves fall dead — a complaint that accompanied Goldberg throughout her life. Moreover, she hints that the two are connected, that the loves die because of the poems. Isn't it that the birds smash against the windows?

Sincerity by itself, however, does not produce beauty, and if the message were presented directly, the poem would not be as moving. The last lines, which deliver a blow to the stomach, penetrate our armor mainly because we are unprepared. And the poem's way of achieving this is the incidental, offhanded statement of the message. The birds-loves as if serve only to exemplify the transparency of the windows. The dead loves are presented as a mere illustration of something else. This is displacement.

Displacement, like all incidental communication of powerful emotions, has great force. The reader feels as if he were being brushed by a feather, not being sure if it touched him or not, which makes him shiver, as every good poem should do.

Lea Goldberg (1911–1970); Born in Kovno, Lithuania; Immigrated to Israel in 1935.

When a straight line meets a polygon

In mathematics, and in science in general, a change of perspective is often the key to the solution of a problem. The role of displacement here is different from that in poetry. It is not meant to disguise the message, but to cast things in a new light. However, the sensation of beauty is generated the same way. In both mathematics and poetry, the secret is that the message is not completely understood. Things happen too fast. The idea is so new that at first encounter it is not consciously absorbed.

Here is an example. Look at the hexagon in the picture. It is not convex, that is, it has cavities. In the illustration, a straight line meets all six sides.

A six-sided polygon (a hexagon), with a straight line that intersects all its sides. Can you also draw a seven-sided polygon (a heptagon), with a straight line intersecting all its sides?

When a mathematician asks you to perform a task, chances are that he is pulling your leg. It is likely that the mission is impossible. If you try (and I suggest trying to actually feel this hands-on), you will quickly realize that it cannot be done. Why? One way to see this is by change of perspective. The question begins with a polygon, and asks you to construct a straight line that will meet all its sides. Try the opposite: begin with the straight line, and try to draw the polygon.

Before presenting the solution, let me formulate the principle on which it is based. It is called the "river crossing principle." An even number of crossings of a river brings you back to the original bank, and an odd number of crossings takes you to the other side.

By this rule I could know, for example, whether I went through the door of my office an even or odd number of times (my office is located on the sixth floor, and I can't enter or leave it through the window). I don't know what this number is, but I am certain that it is even: every time that I entered, I also went out (these lines are being written outside the office). As simple as this principle may appear, it is at the heart of many profound mathematical theorems.

So, let us draw the straight line, and then try to draw the heptagon. Let us start at point Q and proceed along the sides of the broken line. How many times will we cross the straight line on the way? Seven times, of course, since each of the seven sides crosses the straight line. Since 7 is odd, by the river crossing rule the heptagon must end on the other side of the straight line, not on the side on which it began. That is, it ends on the side opposite Q. But since the heptagon is closed, it should end at Q. This contradiction means that the assumption that every side of the heptagon intersects the straight line is impossible.

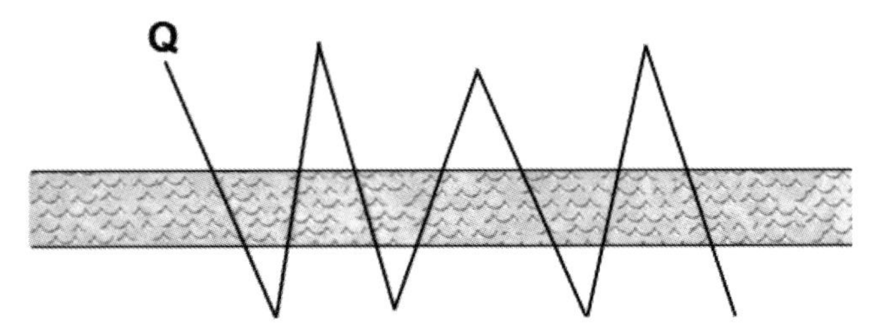

After seven crossings, we are on the other side of the river. The polygon is not closed.

How Many Games Are There in a Knockout Tournament?

A knockout tournament is a competition in which the players are paired off. Each pair competes and the winner advances to the next round.

Question: How many games will be played in a tournament with 16 players?

In the first round, the 16 players are arranged in 8 pairs, 8 games are played, and the 8 winning players go on to the next round. These 8 players are arranged in 4 pairs, and will play 4 games. In the third round, there will be 2 games, and in the fourth, that will determine the champion, only 1 game. The total number of games is therefore $8 + 4 + 2 + 1 = 15$.

The number 16 is a power of 2 — it is 2^4, that is, $2 \times 2 \times 2 \times 2$. As a result, in each round all the players can be paired off. But a tournament can also be held with a number of players that is not a power of 2. In such a case, in some rounds there will be an odd number of players, and they cannot all play. When this happens, the players are paired off except for a single player, and the extra player advances to the next round without playing. How many games will be held then?

A secret shared between mathematicians and poets is thinking in concrete examples. Mathematicians also know that the simpler the example is, the better. There is no such thing as a "too simple example." The simplest example here is that of a single player. In this case, the number of games is 0. In the next simplest example, a competition with 2 players, there is a single game. When there are 3 players, 2 games are held: a game between a pair, followed by game between the winner of the first round and the player who was waiting on the sidelines. Let us now skip to a tournament with 10 players. The following diagram depicts a possible course of events. In the first round 5 games were played; in the second, 2; in the third, 1; and in the fourth, again 1.

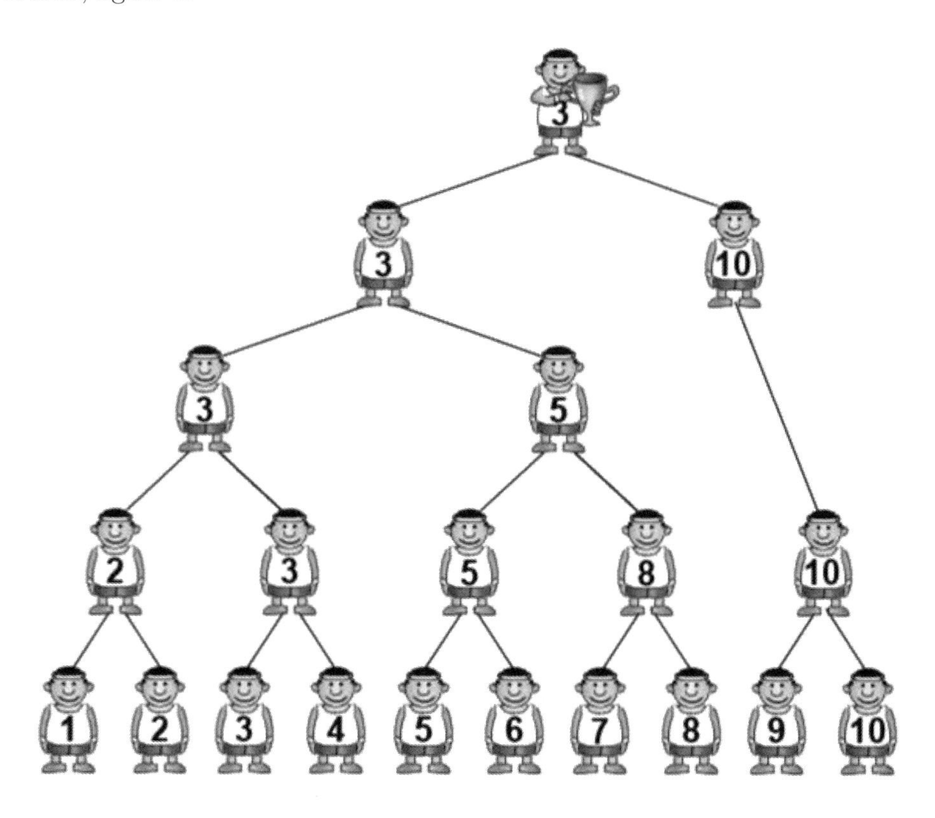

Together: $5 + 2 + 1 + 1 = 9$. This is not very hard, but in the case of 1000 players, the calculation will be tiring. Is there an easier way? Note that in all cases we met, the number of games equaled the number of players, minus 1. Is this coincidental? Almost certainly not. It is probably a rule: **the number of games is smaller by 1 than the number of players**.

A good conjecture is essential, but it requires proof. And the proof here is done by change of perspective. Instead of looking at the winners, look at the losers. When there are 1000 players, each of the 999 players who did not win the cup lost exactly once: there was exactly one game in which he dropped out. Since each game has exactly one loser, for 999 players to lose, there must be 999 games.

The solution leaves us with a sense of beauty. It was economical — it saved effort. It was concise, like a good poem. And there was magic. Things happened fast, too fast for us to fully comprehend at first encounter. Like in Lea Goldberg's poem, the message was slipped under our noses, our attention having been drawn to something else. Judging by the examples so far, a sense of beauty is born in the unconscious grasp of a notion. We experience beauty when we receive a strong message, whether emotional or intellectual, while comprehending it only subliminally.

Part I: Order

God is a mathematician of a very high order

James Jeans, English physicist, 1877–1946

The Curious Case of the Ants on the Pole

Only about myself did I know how to speak.
My world is as narrow as that of an ant.

Rachel Bluwstein, Israeli poetess, 1890–1931

There is an unknown number of ants on a one-meter long pole. The ants move — some to the right, others to the left, but all at the same speed: exactly one meter a minute. The pole is narrow, about as wide as a single ant, and when two ants meet they cannot continue. They then behave like colliding billiard balls, that is, each turns about and continues in the opposite direction, at the same speed.

When two ants meet (left) they change direction (right).

Every so often, an ant reaches one of the pole ends, and then it falls off and disappears forever.

Question: In the end, will all the ants fall off the pole? If so, how long will this take?

At first glance, the answer seems to depend on the initial state, that is, on the number of ants on the pole and their position. If there are many ants, it seems that it might take a long time for all of them to fall off. How can we test this? I have already told you the first secret of thinking mathematically: studying examples. Mathematical thought is a play between examples and abstractions. The difference between the two is that strokes in the direction of the concrete can be done consciously, that is, examples can be evoked deliberately. For this reason one ought to begin with examples. An additional reason, of course, is that examples are the raw material of abstraction. In the case of the ants, the simplest example is that of a single ant. If the ant is at one end of the pole and goes toward the other end, it will fall off in one minute. In any other case, it will fall in less than a minute. But we still have not touched upon the core of the problem: the collisions. So let us look at two ants, located at the opposite ends of the pole, advancing toward each other.

After half a minute, they will meet in the middle of the pole, reverse their directions, and fall off in another half a minute. So, both will fall after exactly one minute.

The next example is a bit less obvious. Imagine one ant starting at the right end, the other exactly in the middle, and they are advancing toward each other.

The ants will meet after a quarter of a minute at a distance of a quarter of a meter from the right end. They will reverse their directions, and then the last ant to fall will be the ant on the left, that will fall after three quarters of a minute. After exactly one minute, both ants will have fallen.

This is starting to look strange. In all three examples, all ants fell off the pole within one minute. Let us go up one level of complexity higher, and examine three ants. Consider the case where ant A starts from the left end, and moves to the right; ant B starts from the middle and moves to the right; and ant C starts at the right end and moves to the left.

After a quarter of a minute we shall see the following picture:

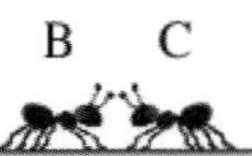

After the collision, ants A and B will go towards each other, meet in the middle after another quarter of a minute, reverse their directions and each will fall after another half a minute. Ant C will had fallen from the right end even before that. Again, it will take a total of one minute for all ants to fall.

Now this is really strange. In all our examples, all the ants fell within a minute. Does this always hold true? The answer is "yes," and the proof is easy. That is, if you have the right insight. Strange as it may seem, this insight does not add information but ignores information: it ignores the identity of the ants. If we don't care who the ants are, then what happens at the moment two ants meet? Actually, nothing. Before their meeting, one ant goes to the left, and the other to the right; after their encounter, the exact same thing happens: then too, one ant proceeds to the left, and the other to the right, at the same speed. But for our purpose, which ant goes to the left and which goes to the right doesn't matter.

So in effect, there are no collisions. They were only there to confuse us. The problem is completely identical to the problem: "ants are proceeding along a one meter long pole, each at the speed of one meter per minute, without colliding and without changing direction. How long will it take for them to fall?" There is no mystery here. All will fall off in one minute or less.

Mathematicians are a lucky breed. They get paid to play. When we take into account the billions that are invested in mathematical research and education, we would expect them to be busy with applied projects. In reality, most mathematician allow themselves to indulge in problems like this one. Why? Because the impractical appearance of this riddle is misleading. In fact, it is a good example of the discipline's primary strength: abstraction. The ants in the problem are mathematical: real ants do not move at a uniform speed, and do not obey such simple rules. Mathematics is the study of systems that follow well defined rules. And the abstraction is even more evident in the solution, that strips the situation of its details, and exposes its essence.

Ignoring the irrelevant, as in the ants problem, is a primary characteristic of mathematical thought. Mathematics takes the abstraction process to its extreme. It takes a complex looking tree, strips it of its leaves, and reveals the trunk. Think, for example, of the concept of number. The person who invented the number "4" understood that, as far as the rules of arithmetic are concerned, it is immaterial if he had 4 stones or 4 pencils, what color they are, and how they are arranged. 4 stones and 3 stones are 7 stones, just as 4 pencils and 3 pencils are 7 pencils, and hence we can say abstractly "$4+3=7$." Abstraction is generalization, and generalization saves effort. The rule we found for stones will be valid for any kind of objects, and at any point in time. "Mathematics is being lazy," said the mathematician George Polya (1887–1985), "it is letting the principles do the work for you." In this respect, the ants question is very practical. Directly, it is not useful for anything, because there are no ants like these in reality, but it educates the person who solves it to think abstractly.

It may even be the case that problem was invented to model a real-world phenomenon. Bundles of light waves ("solitons") behave in collisions just like the ants in the solution: they pass through one another.

Hidden Order

Nature does nothing in vain, and more is vain, when less will serve; for Nature is pleased with simplicity.

Isaac Newton

Isaac Newton, English mathematician and physicist (1642–1727). In 1666, his "miraculous year," he fled from the plague to the village where he had been born, where in one summer he developed the theory of gravity, several of the principles of modern optics, and differential and integral calculus. He spent his later years in disputes over priorities (especially with the German Gottfried Wilhelm Leibniz, on the discovery of differential and integral calculus), in experiments in alchemy, and as master of the British royal mint.

The power of concepts

A good concept is like a path that suddenly opens before you in a dark forest. A minute ago the thicket seemed impenetrable; but from the moment the path was revealed, the way stands open. The English mathematician Andrew Wiles, who solved the famous Fermat's Conjecture, used another metaphor. A good concept is like a light switch, that you find as you feel your way in a dark castle. When you turn on the light, you know what is in the room you are in. In the next room you will have to look for another switch.

Here is a classic example, a problem composed in 1946 by Max Black, a British philosopher and mathematician. Take an 8×8 board of 64 squares, and cut out the lower left-hand corner and the top right-hand corner, like this:

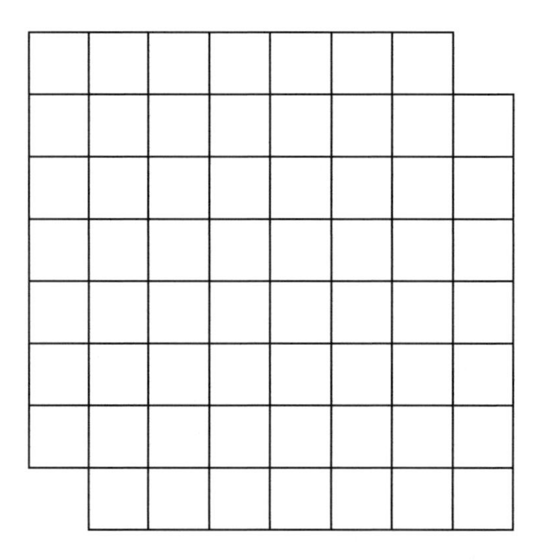

Can all 62 squares be covered by 31 dominoes?

You also have 31 dominoes, each of which can cover two adjacent squares of the board. All together, they can cover 62 squares, which is the number of squares on the incised board. Can the board be covered with these dominoes?

You may have guessed the first step: look at small cases, even very small. The smallest possible example is a 2×2 board. After the removal of two opposing squares, we get:

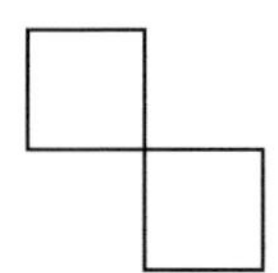

Of course, this shape cannot be covered with a single domino. Now try a 4×4 board (We are skipping the case of a 3×3 board, since it contains 9 squares, and removing 2 leaves 7 squares, which is an odd number. An odd number of squares cannot be covered without overlapping, since each domino covers 2 squares). A bit of experimentation will convince you that this is impossible.

In a 2×2 or 4×4 board it is easy to check all possibilities. In an 8×8 board this would be impractical, there are too many possibilities. We need an idea. And the concept that hits the mark is coloring the squares black and white.

Things now fall into place. Each of the 31 dominoes will cover one black square and one white square. Since the squares we removed from the board are both white, there are 32 black squares left, and only 30 white squares. These cannot be covered by 31 dominoes that are supposed to cover 31 squares of each color.

The chessboard coloring revealed a concealed pattern. Emerging by magic, as if from nowhere, it made things simple and clear.

The chocolate problem

Here is another example for the power of concepts. A chocolate bar measures 5 squares long and 4 wide. We want to divide the 20 squares among

20 children, so we must break the bar into its squares. The rule is that at each step we may take one of the pieces we have at hand, and break it along a single straight line. What strategy should we follow to apply as few breakings as possible?

As usual, the first step is to consider simpler examples, which in this case means having smaller dimensions. For example, a 3×1 chocolate bar divided among 3 children. Here we have no choice: 2 breakings are necessary. Let us move on to a slightly bigger example: a bar of size 3×2, to be divided among 6 children. One way would be to first separate the two rows of 3 blocks apiece by a single breaking. After this, we need another 2 breakings for each of the 2 rows, for a total of 5 breakings.

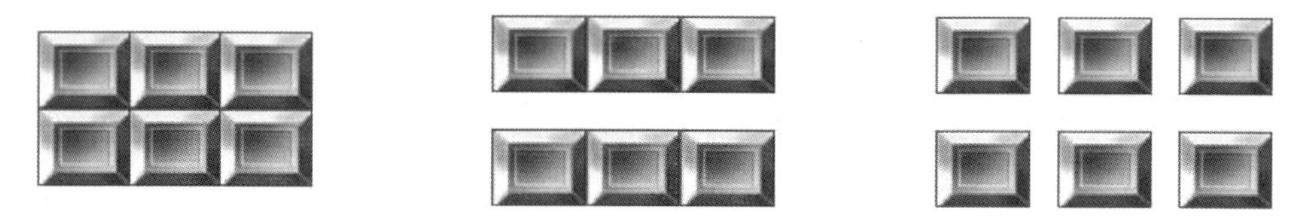

One lengthwise breaking, and 2 more in each row, for a total of 5 breakings.

Another way would be to separate the bar into 3 columns of 2 blocks apiece, by 2 breakings. In each of these 3 columns we need an additional 1 breaking, for a total of 2 + 3 breakings. In this way, we need 5 breakings.

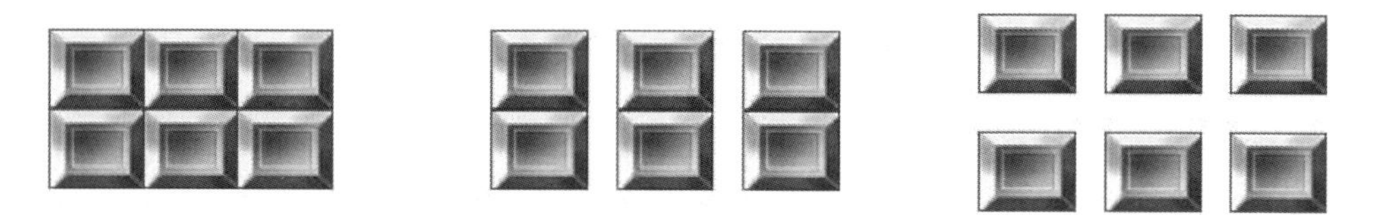

Two vertical breakings, and 3 horizontal ones, for a total of 5 breakings.

The different ways lead to the same result: **a chocolate bar of n blocks needs $n-1$ breakings.** Namely, the strategy has no effect on the number of breakings. We cannot separate the individual blocks with fewer than $n-1$ breakings, nor with more than $n-1$. Why is this so? Here, again, a correct concept makes things simple. This is **the number of pieces obtained after each breaking**. At the start the number of pieces is 1 — there is a single bar. Each breaking turns one piece into two, thereby increasing the number of pieces by 1. At the end of this process, there are n pieces. In order to go from 1 to n pieces, with each step adding a single piece, we need $n-1$ steps.

As usual, once we have found the correct concept, we can generalize. It transpires that the requirement of breaking along straight lines is irrelevant. We can break any way we want, as long as each breaking turns a single piece into two pieces.

To Discover or to Invent

Platonism

Don't go around saying the world owes you a living; the world owes you nothing; it was here first.

Mark Twain, American writer and humorist, 1835–1910

A standard physics department has both theoreticians and experimentalists. Experiments are supposed to be the raw material for theories. In a mathematics department it is rare to find experimentalists (surprisingly enough, there are exceptions: in a leading Canadian university there was a Laboratory for Experimental Mathematics for some period of time). Mathematicians don't need laboratories. They do their work in an office, on a blackboard or on paper, and all they need is their minds. In this aspect they are superseded only by philosophers.

> The president of the university visited the mathematics department.
> "You know," he told his hosts, "of all faculty members, I like mathematicians best. All they need is paper, a pen, and a wastebasket."
> Ruminating, he then added, "Philosophers are even better. They don't need the wastebasket."

Scientists study the world. What do mathematicians investigate? Is it something that exists in the world or the products of their feverish minds? In short, do mathematicians discover or invent? Does mathematics discover order that exists in the world or does it create the order? Do mathematicians construct something new, like a house, or discover something that already existed, the way Columbus discovered America? Is a concept like "even number" part of the external world, or is it only in the mind of the thinker?

This is supposed to be the subject of fervent dispute among mathematicians. The approach that says that mathematical objects are as real as chairs and tables is called "Platonism" (incidentally, the original Platonism is more extreme. Plato argued that the concept of the table is more real than the table itself). A bitter row is supposed to exist between Platonists and anti-Platonists. In practice this is not the case. The twentieth-century American mathematician Ralph Boas claimed that he had never met a mathematician who was not a Platonist. Almost all mathematicians believe in the reality of their objects. Numbers, geometric shapes, functions, evenness of numbers — these are all part of the actual world. Mathematics is discovery, not invention. Mathematics reveals order that is out there in the world. A concept is nothing more than mirror image in our brain of a pattern in reality. A mathematician is more of a photographer than a sculptor.

The sum of an arithmetic sequence

Here is a discovery of order in the world — one made by a seven-year old. This is one of the best known stories in the history of mathematics. Carl Friedrich Gauss (1777–1855) was the greatest mathematician of the nineteenth century, some say of all times. At the age of three, he corrected a mistake in calculations made by his father, a bricklayer. When he was seven, the teacher of the class he attended wanted to have an hour's rest, so he told his students to calculate the sum of the numbers from 1 to 100. To his surprise, Gauss came to him after a few minutes with the answer — 5050.

How did little Carl Friedrich do this? By finding order. In the sum $1+2+3+\cdots+98+99+100$, he matched 1 with 100, 2 with 99, 3 with 98, and so on. The sum of each pair is 101. Since there are 50 pairs, the total sum is 50 times 101, which is 5050.

The number 100 is even, so the terms between 1 and 100 could be matched in pairs. How can we sum up the number between, say 1 and 1001? One trick is adding 0 at the beginning. This, of course, doesn't change the sum, and now the number are paired off: 0 with 1001, 1 with 1000, and so on. After the addition of 0 there are 501 pairs, each summing up to 1001, so the total sum is $501 \times 1001 = 501{,}501$. Another way, perhaps the most direct, is this: the average size of a number in the sequence $1,2,\ldots,1001$ is the middle between the first element and the last, namely between 1 and 1001, which is 501. The sum of 1001 numbers is 1001 times their average, again 501×1001.

Was this Gauss's invention, or discovery? Obviously, discovery. And in fact he was not the first. This was discovered before him, and re-discovered after him. Mathematical ideas are there to be discovered. If one mathematician misses the idea, another will find it. That is why, in my opinion, the tragically early (age 31) death of Franz Schubert was a greater loss to humanity than the even earlier age death of Evariste Galois, the French mathematical genius, at around the same time — early 19th century. The discoveries that would have been made by Galois, had he lived to an old age, have long since been made, while with the death of Schubert we lost unimaginable treasures of beauty.

Carl Friedrich Gauss (Germany, 1777–1855), the greatest mathematician of the 19th century. He contributed to the theory of complex numbers, number theory and modern algebra. Together with the physicist Wilhelm Eduard Weber he built the first telegraph. He spent his later years in seclusion in the observatory in Gottingen, and published very little. The biographer Eric Bell estimated that if all of his discoveries had been published in his lifetime, mathematics would have progressed by fifty years.

Is poetry invented or discovered?

And what about poetry? Is it in the world, or in the poet's mind? We would think that the answer is clear: obviously, poetry is invented. But listen to what a mathematician (and a poet) has to say about this. Sofia Kovalevskaya (1850–1891) was the favorite student of Karl Weierstrass, one of the important mathematicians of the late nineteenth century. In one of her letters she related to Weierstrass's statement that a true mathematician must be something of a poet:

> *In order to understand this, one must renounce the ancient prejudice that a poet must invent something that does not exist, that*

> *imagination and invention are identical [...] The poet must see what others do not see, must look deeper than others look.*

Her words strike true. As we already mentioned, poets, like mathematicians, are hunters of hidden patterns. An on-target metaphor reveals a similarity that is out there. The target had existed before. When the poet Yehuda Amichai writes,

> *Careful angels threaded fate within fate,*
> *Their hands shook not, nothing dropped or fell.*

Yehuda Amichai, "Twenty New Rubaiyat," *Poems*

He expresses an existing truth: our fate is no more in our hands than the thread is master of its fate; there are forces that direct it, as the seamstress directs the thread. This is beautiful not because it is an invention but, mainly, because it is true. As Franz Kafka claimed, poetry is always a search for the truth.

Order and Beauty

Saving energy

When everything falls into its proper place, we say "Everything worked out beautifully." Why? Recognizing order is useful. It saves effort in coping with the world. But why should it cause aesthetic pleasure?

To answer this, we must first realize that it is not mere order. Order by itself is not necessarily beautiful. Nothing is more orderly than a blank sheet of paper, and no combination of sounds is more orderly than absolute silence. Nonetheless, a blank sheet of paper is not a work of art, and silence does not possess the beauty of a Mozart symphony. A monotone series of beats is orderly and predictable, but it does not constitute music. In order to create a sensation of beauty, we need something beyond order.

The secret lies in a concept proposed in the second half of the 19th century: saving mental energy. The industrial revolution in England led to the idea that machines can replace not only muscle work, but also mental. This led to the invention of the first computer, by Charles Babbage, and also to a mechanical perception of the human mind. One proponent of this was Herbert Spencer, who claimed that the mind, like other systems in the world, seeks a state of minimal energy. Young Freud adopted this approach wholeheartedly, and in the 1890s, when he was still taking first tentative steps in psychoanalysis, he wrote a draft of a thick book entitled *Physiology for Psychologists*, in which he tried to explain mental phenomena in terms borrowed from the physical world of his time. Freud championed the Spencerian idea that the psyche tries to reduce effort as much as possible, that is, to save energy.

Like many before and after him, Freud quickly learned that psychological terms that are effective as metaphors soon become useless when used concretely. The concept of "saving energy" is too general to predict the behavior of human beings. As a result, *Physiology for Psychologists* was shelved around the year 1895, but echoes of it would reverberate throughout Freud's writings. The idea of saving energy was expressed most clearly in a book he wrote in 1905 on humor, *Jokes and Their Relation to the Unconscious.* The book's thesis was that the pleasure we derive from a joke results from saving the energy of repression. The joke enables us to enjoy forbidden things without having to repress them. Consequently, energy that was prepared to repress the forbidden idea is unnecessary, and is transformed into pleasure.

Not much revelation came out of this book, as far as humor is concerned. Freud himself was not happy with it, referring to it in later years as a needless deviation from his main course. But the idea of saving energy caught on, especially with respect to art. The idea is that a work of art disguises itself as chaotic, demanding preparation of energy to tackle it, and then hidden order is revealed, which means that the energy prepared can be saved. And saved energy entails pleasure. Just like when we discover that we won a battle we feel pleasure, because we no longer need the energy we prepared for the struggle. The sensation of beauty, by this approach, arises when order is suddenly revealed in disorder.

Music is one area in which this explanation works beautifully. In order for music to be enjoyable, it has to be complex. It must seem to be disorganized noise, and then to be realized as ordered. We constantly attempt to decipher the stimuli that arrive from the outside world, and so we prepare energy to organize noise. If we then discover order in the noise, this energy is saved. Links between the sounds are revealed, which enable us to predict what is coming. This happens in two dimensions: rhythm and harmony. Rhythm is the organization in time, and harmony the connection between the frequencies of the notes. In the next chapter, I will explain a bit about both.

If the music is complex enough, these links are not straightforward, and cannot be perceived consciously. This means that on the conscious level we do not fully understand the order in the musical work. There is a gap between the perceived lack of order and the hidden order that is unconsciously revealed. And this gap, between what we consciously observe and the unconscious perception, is the source of beauty.

Mathematical Harmonies

Rhythm and prediction

The pleasure we obtain from music comes
from counting, but counting unconsciously.
Music is nothing but unconscious arithmetic.

Gottfried Wilhelm Von Leibniz, German mathematician
and philosopher, 1646–1716

Like all animals, humans are future-oriented. They look where they are going to be in a moment, not where they were. Even a historian, when he prepares an omelet, is more interested in where the egg will be in a moment than in where it was. There's a simple reason for this: this is how living creatures were formed by evolution. Evolution selected those life forms that are best at leaving descendants after them.

Understanding the order in the world means being able to mould your future surroundings to your advantage. This is why we derive pleasure from musical rhythm. An expected rhythm saves the investment of energy in the deciphering of the order hidden in the sounds. But it must not be too predictable, because in order to save energy, energy must first be rallied. If the rhythm is sufficiently complex, and we are incapable of consciously deciphering it, we prepare energy in order to guess the next note. When the order is revealed, this energy is no longer necessary — we know what to expect. The saved energy then turns into pleasure.

Pythagoras

And what about the second element of music, harmony? This is more of a puzzle. We all know that some combinations of notes are pleasing to the ear, while others are less so. For example, a C note sounds well with the C one octave higher. Actually, when hearing them together we can hardly distinguish between them. The C-G and C-E combinations, as well, sound well together. The notes C, E, G are the basic chords of the C major scale, the scale whose notes are played on the white piano keys. A composition in C major frequently begins with the notes C, E, G in some order, strays and wanders about, before finally returning to them. Music is built on the tension between the digressions and the original harmony.

But why is one combination of notes pleasing, while another grates on our ears? Surprisingly enough, the answer to this question is mathematical, and it was discovered by one of the most fascinating figures in the history of mathematics, Pythagoras. He was the founder and leader of a most rare entity: a mathematical cult. The cult numbered about 600 men and women, who lived in the Greek colony of Crotona in the south of the Apennine peninsula, in the "heel" of the Italian boot. They donated all their possessions to the community, and swore to keep their discoveries secret. Legend has it that, one day Pythagoras was passing by a blacksmith's workshop, and realized that when the blacksmith struck rods with a simple ratio between their lengths — for example, one was twice as long as the other or one and a half times as long, the combination of the two sounds was pleasing to the ear.

In modern terminology, two sounds sound well together if the ratio between their frequencies is simple, that is, it is expressed by small numbers (for example 3:2 is simpler than 11:5). The frequency of a sound is the number of times per second the air vibrates when the sound is produced, or in more precise language: the number of peaks per second of the sound waves. If the note is produced by a string, this is the number of vibrations per second of the string. A difference of a single octave between notes (like that between a C and the C above) means a ratio of 2 between their frequencies: the frequency of a high C is twice that of the C below. The frequency of the note G, the fifth in the octave (when beginning with C) is $3/2$ times that of the low C of that octave. In other words, for every 2 vibrations of the C, there are 3 vibrations of the G. The ratio between the frequencies of E and C is 5:4, again quite simple. This is why C, E, and G sound well together.

Helmholtz

Why do simple ratios between frequencies cause pleasure? Pythagoras discovered the phenomenon, but was unable to explain it. Another 2,400 years

would have to pass before this question could be answered. The enigma was solved by the German Hermann von Helmholtz (1821–1894), a true Renaissance man: a mathematician, physicist and physiologist, who also studied aesthetics. His explanation was based on the phenomenon of "overtones." When a chord vibrates at a certain frequency, it also vibrates, at the same time, at frequencies 2, 3, 4,... times higher. The overtones are weaker the further they are from the original tone, namely the higher the ratio is to the original frequency, but they are audible. In other words, when the note C is played, most times we will also hear the C of an octave higher, with a frequency exactly double, and also the G in the higher octave, whose frequency is three times that of the original C. Simple ratio between two frequencies means that they share overtones. For example: the C and the G in the same octave share the G of one octave higher. Hearing these notes together, we reveal hidden order. The notes are different, but, unconsciously we find a factor common to both. Instead of chaos, order emerges.

Does this explain in full the pleasure people derive from music? Of course not. It does not explain how come music can be so moving. It does not touch upon the emotions aroused by music. It only relates to a pleasure that can be classified as intellectual. But it is a good first start.

Mystical numbers

All this was beyond the knowledge of the ancient Greeks, who knew nothing of frequencies. When people don't know, they fantasize. In order to explain harmony, Pythagoras and his school invented fanciful theories of the magical powers of numbers and the ratios between them. "All is number" was their strange slogan. That is, the world is ruled by simple numerical ratios. The Pythagoreans believed that every important natural phenomenon has to obey numerical laws. They maintained that there are simple ratios between the diameters of the planetary orbits, and that the planets consequently emanate "celestial music." And they went far beyond that. They claimed that every size in the world that is of any significance can be expressed as a ratio between whole numbers.

A number that is the quotient of two whole numbers is called a "rational number" (from the word "ratio"). Every whole number is rational: 4, for example, is rational because it is the ratio between itself and 1, that is, $4{:}1 = 4$. Every fraction with a whole-number numerator and denominator is rational, because the fraction bar is actually a division sign: $17/3$ is the ratio between 17 and 3. So, the Pythagoreans believed that important quantities in nature are rational.

Sobering up

The intellectual achievements of the ancient Greeks were nothing short of miraculous. A very small people, numbering no more than a few hundred thousand, developed conceptual systems whose fruits we continue to enjoy to the present. They were motivated by their infinite respect for abstract ideas. For the Greeks, abstractions held magical power, and were more important than the real world. The Greeks were the first to study abstract concepts for their own sake, without regard for applications. The Egyptians and the

Babylonians studied numbers before, but they did so for practical ends. The Greeks were the first to see numbers as a world worthy to be explored for its beauty and inner harmony.

But even within the Greeks' achievements, geometry enjoys a special pride of place. It was in this field that the Greeks developed the concepts of "axiom" and "proof," and it was here that they reached the highest level of abstraction. Pythagoras was one of the founders of Greek geometry. The theorem that, to this day, is regarded (and rightly so) as the most important and useful geometric theorem, is named after him, though in fact he was not its discoverer. The theorem states that the sum of the areas of the two squares based on the legs of a right triangle equals the area of the square based on the hypotenuse. This is important because it enables us to calculate distances. Given the lengths of the legs of a right triangle, we can calculate the length of the hypotenuse. This means that knowing how to calculate east-west and north-south distances, you can calculate the distance between any two points.

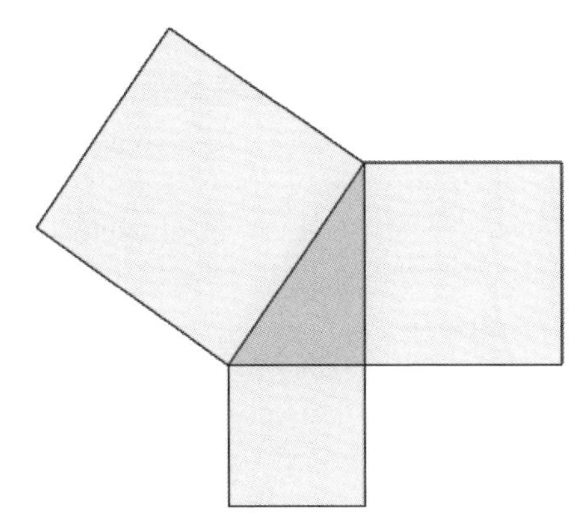

The Pythagorean theorem: the sum of the areas of the two squares on the legs of a right triangle equals the area of the square on the hypotenuse.

An interesting special case of this theorem is that in which the legs of a right triangle are of the same length. Look at a square whose side is 1 unit long. The diagonal of the square is the hypotenuse of a right triangle, with equal legs, each 1 unit long.

According to the Pythagorean theorem, the diagonal, squared, is equal to $1^2 + 1^2 = 2$. Therefore, the length of the diagonal, not squared, is $\sqrt{2}$. There is a simple and especially beautiful proof for this individual case of the theorem, that appears in a somewhat surprising place: one of Plato's

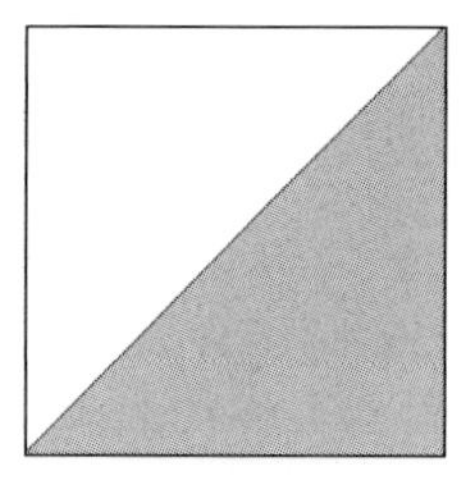

According to the Pythagorean theorem, the length of the diagonal of a square 1 unit long is $\sqrt{2}$.

dialogues (in which, as in all of Plato's dialogues, the hero is Socrates), entitled *Menon.* Look at the next drawing:

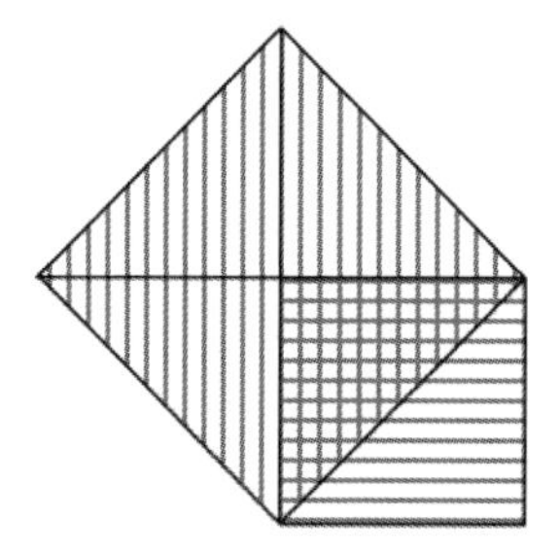

The area of the square based on the diagonal (with vertical lines) is twice the area of the small square (with horizontal lines), because it contains 4 triangles, while the small square contains only 2 triangles. The side of the larger square is therefore $\sqrt{2}$ the length of the side of the small square.

Let's assume that the length of the side of the small square (with horizontal lines) is 1. The area of this square is therefore 1×1, that is, 1. The large square, that is on a diagonal (and marked with vertical lines), is composed of 4 triangles while the small square contains only 2 (all of the triangles are congruent, that is, they are capable of perfectly fitting one over the other). Therefore, the area of the large square is twice that of the small square, which means that it is 2. The length of any square's side is the square root of its area, and so the length of the large square's side is $\sqrt{2}$. But look: the side of the large square is the diagonal of the small square! Therefore, this diagonal is $\sqrt{2}$ long.

Geometry had a special place in the minds and hearts of the Pythagoreans, and for them the diagonal of a square was an everyday object. Therefore, they believed that the length of this diagonal should be rational. For many years they tried to find what ratio it is. It is close to $\frac{7}{5}$, but it is not quite that, because the square of $\frac{7}{5}$ is $\frac{49}{50}$, which is almost 2, but not quite.

Eventually they had to realize the bitter truth: that $\sqrt{2}$ is not rational. This was such a severe blow that they vowed to keep it a secret. Due to the sect's secretiveness, not much is known about it for certain and the continuation of the story might very well be spurious. But legend has it that Hippasus, a sect member who revealed the secret to the world, was put to death for doing so. This is almost certainly apocryphal. Hippassus drowned, and his death might very well have been an accident. But the sect accredited it to punishment by the gods.

Why $\sqrt{2}$ is Not a Rational Number

Actually, why isn't $\sqrt{2}$ a rational number? Why can't it be expressed as $\frac{m}{n}$ for some whole numbers m and n? To see this, assume that $\frac{m}{n} = \sqrt{2}$. We shall show that this assumption leads to a contradiction. First, we can assume that $\frac{m}{n}$ is a reduced fraction, namely the numerator and the denominator are not divisible by the same number, greater than 1. If not, we just reduce it, namely divide by the common divisor. By the definition of $\sqrt{2}$, the fact that $\frac{m}{n} = \sqrt{2}$ means that $\frac{m^2}{n^2} = 2$. If we multiply both sides by n^2, we get:

$$(*) \qquad 2n^2 = m^2$$

Now, we will divide our discussion into two cases: in one case, m is an odd number, and in the other case, it is even. Each of these two cases will lead to the desired contradiction. If m is an odd number, the right side of (*) is the square of an odd number, so it is an odd number (the product of two odd numbers is odd), while the left side is a multiple of 2, and is therefore an even number. Since an odd number cannot be equal to an even one, the right side cannot be equal to the left. (As an example of this case, if $\sqrt{2} = \frac{7}{5}$, then $2 = \frac{7^2}{5^2}$, meaning that $2 \times 5^2 = 7^2$. Then the left side, 50, is even, while the right side, 49, is odd.)

Assume next that m is even. Since, by assumption, $\frac{m}{n}$ is reduced, n must be odd (if it were even, then the entire fraction could be reduced by 2). Since an odd number squared is odd, on the left side of (*) there is the product obtained by multiplying 2 by an odd number. Such a product cannot be divisible by 4. But on the right side we have an even number squared, which is divisible by 4. So, the two sides must be different, again a contradiction to the equality sign.

How do we then express $\sqrt{2}$? One way is as an infinite decimal fraction: $\sqrt{2} = 1.4142135623\ldots$, which means that $\sqrt{2}$ can be approximated by the rational numbers 1, 1.4, 1.41, 1.414, and so on. Note that $1.4 = \frac{7}{5}$ and $1.4285714285\ldots = \frac{10}{7}$. $\sqrt{2}$ is almost exactly in the middle between them! Almost, but of course not exactly: the middle is a rational number, and $\sqrt{2}$ is not rational.

The Real Numbers

It may take a long time to realize the full importance of a discovery. With hindsight, the discovery of the existence of irrational numbers was a turning point in the history of mathematics. If $\sqrt{2}$ is not the quotient of two integers, then just how can we describe it? As we saw, the usual way is as an infinite decimal fraction, which is the limit of an infinite sequence of numbers, whose squares get closer and closer to 2. This was the gateway to the fundamental concept of the limit, the cornerstone of the infinitesimal calculus.

The square root of 2 was not alone for long. It was soon joined by additional irrational numbers. The Greeks realized that if the root of a whole number is not itself an integer (another word for "whole number"), then it is irrational. The roots $\sqrt{4}$, which is 2, or $\sqrt{9}$, which is 3, are integers, and therefore rational. But $\sqrt{3}$, $\sqrt{5}$, $\sqrt{6}$ and so on, are not. Nor is $\sqrt{2}+1$ rational: if it were rational, then, since $(\sqrt{2}+1)-1=\sqrt{2}$, the number $\sqrt{2}$ would be the difference between two rational numbers, and hence rational itself, which we know it isn't.

The conclusion is that there are infinitely many irrational numbers. In fact, they are so numerous that they are "dense," in the sense that there is an irrational number between any two distinct numbers. The rational numbers look like a sieve, whose holes are the irrational numbers. An even more startling discovery will be made at the end of the nineteenth century by Georg Cantor: that the holes are the majority. There are more irrational numbers than rational numbers. Among numbers, as among humans, rationality is rare.

The rational and the irrational numbers together are called "real numbers." Of course, this is not a proper definition of the term. It is like defining "living creatures" as "humans or nonhumans," which does not tell us what is a "nonhuman living being." The real numbers were defined precisely only at the end of the nineteenth century, which was an era of transition from fuzziness to rigor, mathematical intuitions being supplemented by precise definitions and proofs. In those years the principles of differential and integral calculus were given explicit and accurate definitions, clear-cut axioms were written for the natural numbers, and David Hilbert completed the work that Euclid had left unfinished for 2,000 years: the writing of precise axioms for plane geometry.

Two mathematicians, Richard Dedekind and Georg Cantor, were responsible for the rigorous definition of the real numbers. Their definitions provided the justification for the way in which these numbers had been presented beginning in the sixteenth century, that is, as infinite decimals. The number π, for example, which is the ratio of the circumference of circle to its diameter, is written as $3.145912\ldots$, going on ad infinitum. What this means is that the numbers 3, 3.1, 3.14, $3145, \ldots$ get nearer and nearer to π.

There is something special in the decimal expansion of rational numbers. Everybody knows, for example, that $\frac{1}{3} = 0.333\ldots$, the 3s repeating forever. This is true of every rational number. The decimal expansion of a rational number keeps repeating from some point on, as in $2.4131313\ldots$, in which 13 repeats indefinitely.

Indeed, what is this number, written as a fraction? There is a simple trick that does the job. It uses the fact that surprises many people and exasperates others, that $1 = 0.999\ldots$. In order to understand why this equality is true, we must first understand what $0.999\ldots$ is: it is the limit of the sequence 0.9, 0.99. $0.999, \ldots$. These numbers approach 1 because their distances from 1 are $\frac{1}{10}, \frac{1}{100}, \frac{1}{1000} \ldots$ — numbers that tend to zero.

Knowing that $0.999\ldots = 1$, we can calculate $0.333\ldots$. Since dividing a number by 1 does not change it, we can write: $0.333\ldots = \frac{0.333\ldots}{0.999\ldots}$. In the last quotient, every 3 in the numerator is matched by a 9 in the denominator. Accordingly, the denominator is 3 times as large as the numerator (when I explain this to children, I tell them about two brothers: for every amount of money that one receives, the other receives three times as much. At the end of the day, the second brother will have three times as much money as the first). This means that the fraction is equal to $\frac{1}{3}$.

Let us now look at $2.4131313\ldots$. The number $0.0131313\ldots$ is one-tenth of $0.131313\ldots$. We can write $0.131313\ldots = \frac{0.1313\ldots}{0.9999\ldots} = \frac{13}{99}$. (When the numerator "receives" 13 one-hundredths, the denominator "receives" 99 one-hundredths; when the numerator "receives" 13 one-thousandths, then the denominator receives 99 one-thousandths, and so on. Therefore, the numerator is $\frac{13}{99}$ as big as the denominator.) Summarizing, $2.4131313 = 2.4 + \frac{1}{10} \times \frac{13}{99} = 2\frac{409}{990} = \frac{2389}{990}$, a fraction.

The other direction is also true: every rational number can be written as a recurring decimal fraction. This is proved, simply, by dividing the numerator by the denominator. $\frac{7}{3}$, for example, is the result of dividing 7 by 3, and performing the division we get 2.333.... It isn't difficult to show that, dividing one integer by another the numerals repeat themselves beginning at a certain point.

This implies, for example, that the number 0.101001000100001...is not rational, since it is not recurring. This is also a (somewhat vague) indication that there are more irrational numbers than rational ones: recurrence is a rare phenomenon, and "most" of the decimal fractions are nonrecurring.

The Miracle of Order

I see myself as a child on the beach gathering shells.

Isaac Newton

The most incomprehensible thing about the world is that it is at all comprehensible.

Albert Einstein, 1879–1955

The unreasonable effectiveness of mathematics

A physical theory must possess mathematical beauty.

Paul Dirac, British mathematician and physicist, 1902–1984

Einstein told us that even if we know something about the order that rules the universe, we will never understand why it is there at all. It seems as if nature built for us a castle, whose treasures we reveal bit by bit. And we shall always be like children playing with shells on the shore of the sea, the depths of which we will shall never fathom. But even more surprising than the existence of order is the fact that it is expressed by mathematical formulas. More than that — by the most advanced mathematical theories of the day. The great number theorist Godfrey Harold Hardy (1877–1942) was an avowed pacifist. His main research partner, John Edensor Littlewood, spent World War I developing artillery. This may be why when, towards the end of his life, Hardy summed up his experience as a mathematician in his book *A Mathematician's Apology*, he took comfort in the fact that none of his discoveries had ever had any use, certainly not for military applications. Not much later Hardy's theories played a role in encryption theory, and now they are indirectly applied in parts of computer science.

The annals of mathematics are replete with such examples. Today's esoteric fields are tomorrow's basic scientific tools. A famous example: in the third century BCE Apollonius (262–190 BCE) developed a theory of conic sections — the curves resulting from the intersection of a plane with a conic envelope. These are the circle, the ellipse (of which the circle is a special case), the parabola, and the hyperbola (it was Apollonius who coined all these names).

The four sections of a cone cut by planes (from top to bottom): circle, ellipse, parabola, and hyperbola.

Apollonius' research was totally theoretical, without the slightest practical intent. Almost two thousand years later, the German mathematician, astronomer, and astrologist Johannes Kepler (1571–1630) used conical sections to describe the movement of the planets. The planets travel around the Sun in elliptical orbits (and in a very special case — in a circle), while the trajectory of an object that comes from infinity and is not caught in the Sun's gravitational field, but continues on its way, is usually hyperbolic; parabolic in very singular instances.

Another famous example is the general theory of relativity. When Einstein needed tools to discuss space-time geometry, he asked his mathematician friends, and they told him that just what he needed already existed. The tools had been developed some fifty years earlier by Bernhard Riemann (1826–1866) and others, who never imagined that their discoveries would be of practical use in the not so distant future.

Mathematics, which draws its problems from the real world, abandons reality to advance on tracks originating from within itself — only to discover that it preceded physics by a few decades or, sometimes, by centuries. Group theory, for example, is a field of algebra that seems too abstract to have any practical use. So much so, that in 1910 the Nobel laureate physicist James

Jeans advocated removing it from the curriculum of Princeton University, where he was visiting. Physics students, so he claimed, would never have occasion to use it. Not long afterwards, group theory became one of the most basic tools used by physicists, in particular in the field of elementary particles. Today's promising subatomic theory, string theory, would not have been born without the tools that were developed in recent decades in algebra and topology. Eugene Wigner, a Nobel Prize laureate in physics, wrote an article that became famous, "The Unreasonable Effectiveness of Mathematics in the Natural Sciences." He wrote about his amazement at the fact that theorems proved for purely theoretical purposes quickly become relevant for the real world. Why, of all disciplines, is higher mathematics the proper tool? And how come it is deep and front line mathematics that is needed?

One possible explanation is that "deep" and "elementary" are relative concepts. The deep mathematical theories used by the ancient Greeks appear elementary to a modern-day scientist. What looks deep to us might appear elementary to creatures of higher intelligence. Physicists just use what they have. If mathematicians provided them with better tools, they would have used them.

Order in the world and order in mathematics

There are more things in heaven and earth, Horatio, than are dreamt of in your philosophy.

Hamlet, in *Hamlet* by William Shakespeare, Act I

Yes, but there are also things in philosophy that have never been dreamt of in heaven and earth.

Georg Christoph Lichtenberg, mathematician and satirist, 1742–1799

Mathematics describes the world, but there are also many things in mathematics that have never been dreamt of in the world. From the moment of the invention of a mathematical concept, it has a life of its own. In fact, most mathematical problems do not emerge from real life problems, but from other mathematical problems. Questions gain their right to exist by relating to earlier concepts. But then, the offshoots often return and join the main river. Scientists suddenly realize that they need them, in spite of their purely theoretical appearance.

Beauty and truth

If the solution is not beautiful, I know it is wrong.

Buckminster Fuller, mathematician, architect, and inventor

The ideas chosen by my unconscious are those which reach my consciousness, and I see that they are those which agree with my aesthetic sense.

Jacques Hadamard, French mathematician, 1865–1963, from *An Essay on the Psychology of Invention in the Mathematical Field*

My work always tried to unite the truth with beauty, but when I had to choose one or the other, I usually chose beauty.

Hermann Weyl, German mathematician, 1885–1955

One of the characteristics most peculiar to mathematics is its conjectures. These strange creatures are the forceful drive and the holy grails (there are many of them) of mathematics. Conjectures may survive for hundreds of years before being solved. And strangely enough, most of them are eventually proved, rather than refuted. How do mathematicians have the hunch that a fact should be true? The surprising answer is that the best criterion is aesthetics. Mathematicians believe a conjecture when they feel it is beautiful.

Godfrey Hardy, the English mathematician already mentioned in this book, received a letter in 1913 from a poor Indian clerk named Srinivasa Ramanujan. The letter contained a collection of identities in number theory. Hardy could prove some of them, but many he could not. He believed they were true, because they looked so elegant. Ramanujan himself could not explicitly prove some of them, but merely "dreamt" them. Hardy, who realized that the young Indian was one of the great mathematical geniuses of all time, invited him to England, where the two worked together for a few years. Sadly, Ramanujan could not withstand the English climate and being away from home. His health, which had not been good to start with, rapidly deteriorated. He died in 1920, after having returned to India.

In the process of proving a mathematical conjecture, it frequently looks as if the blanket is too short: if you pull to one side, the other will not be covered. But if the hypothesis is beautiful, the mathematician believes that the deep order behind it will act in his favor, and that he will uncover its

Srinivasa Ramanujan (1887–1920), a mathematical genius who grew wild in India, without formal university education. He was discovered by Hardy and Littlewood, came to England, and worked with Hardy for about four years. He could not adjust to the English climate, and returned to India, where he died at the age of 33.

underlying logic. And if not he himself, then those who come after him. It seems that the goddess of mathematics is on the side of beauty — the more beautiful the conjecture is, the better its chances of being correct. Beauty is the guide to truth because it expresses an unconscious perception of order. When everything falls into place, there must be an intrinsic reason.

Simple Conjectures, Complex Proofs

Proofs really aren't there to show you something is true — they're there to show you why it is true.

Andrew Gleason, American mathematician, born 1921

The strange usefulness of mathematics is a wonder. Another wonder, no less curious, is that this order appears in concentrated pills. Deep order is often revealed in concise formulations. Easily-formulated arguments, that even a child can understand, float, as icebergs on the ocean, within complex mathematical structures. Their proofs, in surprising contrast, sometimes fill hundreds and even thousands of pages. How can simple facts have proofs that are often tens of thousands of times more complicated than the facts themselves, at least in terms of word count?

Let me describe four such cases. Four simply stated conjectures, whose proofs are difficult or still elusive. Each of the four withstood determined attacks, and two are still unproven. Such conjectures are alluring to both professional and amateur mathematicians. Like a gambler who believes that Lady Luck will favor him of all creatures on earth, every mathematician harbors some hope that he will be the lucky one to be favored by the goddess of mathematics. But innocent looking trees may have deep roots. Most of the time, when the proof is finally found, it is not at all straightforward, and requires new and surprising ideas.

Kepler's sphere packing problem

The owner of an orchard wants to pack his oranges in a carton. As usual, we will assume that these are mathematical oranges, that is, perfect spheres all of the same size. We also assume that the carton is much bigger than the oranges. (This condition is meant to ensure that what happens at the edges of the carton will not be of decisive weight. In a more precise formulation of the problem, we take larger and larger cartons, their size tending to infinity.) How should the orchard owner pack the oranges, so that the largest possible number will fit in the carton? This problem has two natural solutions. One is to line the bottom of the carton with oranges, arranged in straight horizontal and vertical rows, like this:

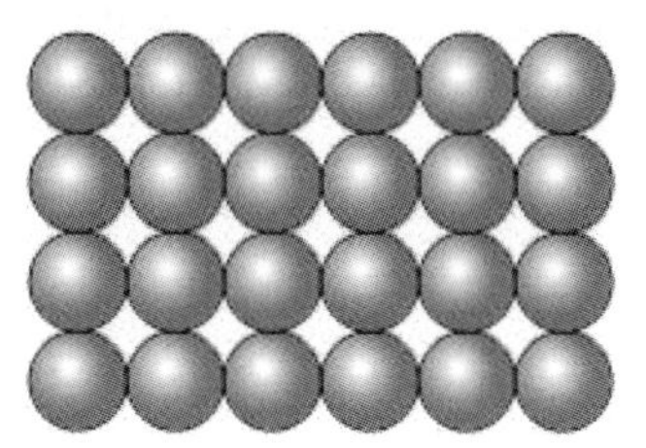

A seemingly economical packing. The next layer will sit in the spaces between the oranges in the first layer.

Over the first layer we will now place the second layer, fitting the oranges into the "holes" between every adjacent quadruple of oranges. The third layer will fit into the holes between quadruples of oranges in the second layer, and so on.

In the second natural solution, the oranges on the bottom of the carton are arranged as a honeycomb, so that each orange is at the center of the six oranges that surround it. Then, the holes between the oranges are filled, as in the previous solution. The second layer, and succeeding layers, all have this same honeycomb structure.

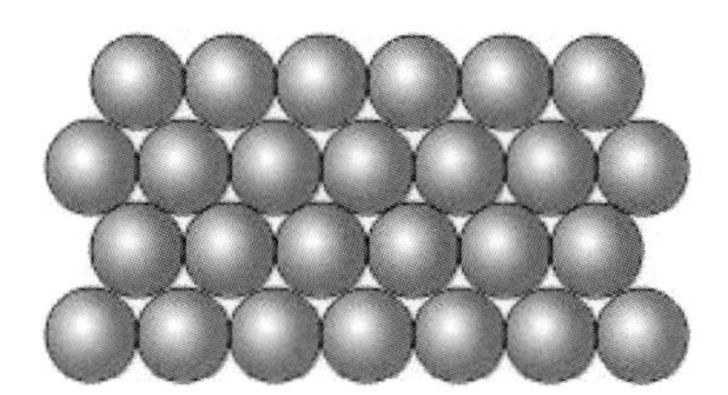

The honeycomb packing also seems economical.

Which of the two packing methods is more efficient? We are in store for a surprise. The two methods, seemingly so different, are actually identical. In the straight rows pattern, there are inclined planes arranged as a honeycomb, and in the honeycomb pattern there are inclined planes arranged in straight rows and columns. This is manifest when we build a pyramid with a square base, with straight rows and columns packing. The illustration below shows that there is a honeycomb packing at the side of the pyramid.

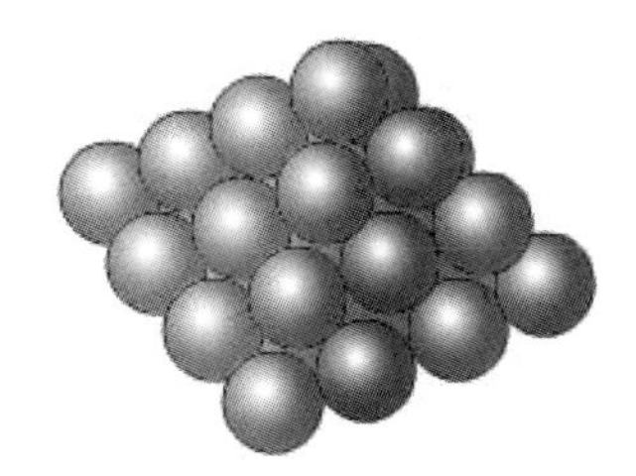

The base of the pyramid is a square, in which the balls are arranged in straight rows in both directions. If we look at the face of the pyramid, we see balls arranged as a hexagon around a central ball — the honeycomb packing.

The fact that the two most natural packing methods coincide suggests that this is indeed the most efficient packing. Johannes Kepler, whose name was already mentioned in connection with the conic sections, surmised that this is indeed the case. This very natural proposition waited 300 years to be proved. Like many other famous conjectures, many incorrect solutions were offered for it over the course of time. A proof that was accepted as correct was found only in 1998 by the American Thomas Hales. Seven more years would pass until the mathematical community agreed on the correctness of the proof. The reason was the proof's extensive use of computers, checking details that are too complex to be done with pen and paper. The length of the written part of the proof is also formidable: some 250 pages!

The four colors theorem

The basic requirement of a political map is that any two adjoining countries are colored differently, in order to be distinguishable from one another. The more colors a mapmaker has at his disposal, the easier it is for him to meet this requirement. For example, if the number of colors equals the number of countries, no special effort is needed — each country has its own color.

In 1852 the English mathematician Francis Guthrie noted that four colors suffice for a proper coloring of the map of England's counties. As a mathematician (or as a poet) this prompted him to generalize. Is it not the case for every map? Can't every map be colored by just four colors? This problem gained immediate publicity, and also an endless number of incorrect solutions. The most famous of these was by Alfred Kempe in 1879. Unlike other solutions, much time would pass before Kempe's error was discovered; in the meantime, partly thanks to his false solution, Kempe was elected a fellow of the British Royal Society. After 11 years, it transpired that he had proved less than he had claimed: that it was possible to color any map in five colors. Election to the Royal Society is for life, and his membership remained in force.

From then until 1976, when the theorem was finally proved, it was the fate of every mathematician in the relevant field, combinatorics — which happens to be my own field — to receive false proofs from amateurs who tried their luck. When a proof was finally discovered, by the Americans Kenneth Appel and Wolfgang Haken, it became apparent to all that there was a good reason for the elusiveness. Not only was the proof long and complex; like Hales' proof for the Kepler theorem, it made extensive use of the computer to check more than a thousand special cases. The proof has been somewhat simplified since then, but until this very day there is yet no proof that does not rely on computers.

I cannot tell you much about the solution, but, as compensation, let me tell you a more modest proposition, whose proof is easy. Assume that the map is drawn in a special way, by adding one circle at a time, such as the left-hand drawing on this page. The circles cut the world into "countries." In this case, you do not need four colors. Two suffice, as in the right-hand drawing:

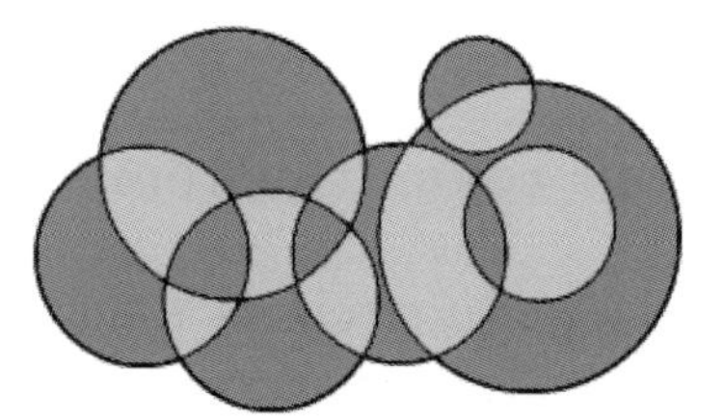

The map on the left is special: The borders are generated by circles. Such a map can be colored with only two colors, as in the example on the right.

The simplest proof of this employs the concept of evenness and oddness (once again, we see how useful this concept is!). Color each country lying within an odd number of circles red, and color each country contained in an even number of circles blue. In particular, the surrounding area (the "sea"), that is contained in zero circles, is colored blue: zero is an even number.

Let us show that this coloration fits the bill. That is, every two adjacent countries are colored differently. Look at one country (call it A). Assume, as an example, that A is contained in 5 circles. Since 5 is an odd number, by our coloration rule, A is colored red. We have to prove that any country (call it B) adjoining A is colored blue. On our map, crossing a border between two countries means entering or leaving a circle. If to go from A to B we leave a circle, then B is located within 4 circles (one fewer than A), and therefore is colored blue (since 4 is even). If to go from A to B we enter a circle, B lies within 6 circles. Since 6 is even, B has to be colored blue, which is just what we had to show. You can easily convince yourselves that there is nothing special about the number 5. The argument is valid for every number.

Ascending and descending sequences

In 1937 the German mathematician Lothar Collatz posed an innocuous looking problem, a mathematical snakes and ladders game. In it, a sequence of numbers is defined according to the following rule: when we get to an even number, we divide it by 2; when we have at hand an odd number, we multiply it by 3 and add 1.

Suppose, for example, that we begin with 10. Since 10 is even, we divide it by 2. The result, 5, is odd, and so we multiply it by 3 and add 1. This yields 16, which is even, and is therefore divided by 2. The sequence we produce is: 10, 5, 16, 8, 4, 2, 1. If we begin with 100, we get:

$$100, 50, 25, 76, 38, 19, 58, 29, 88, 44, 22, 11, 34, 17, 52, 26, 13,\\ 40, 20, 10, 5, 16, 8, 4, 2, 1$$

Both examples ended at 1. Is this always the case? Collatz's conjecture is that this is so. Wherever the sequence starts, it will always end up at 1. Naturally, such a simple looking problem attracted a lot of attention, with no real progress. The Hungarian mathematician Paul Erdős argued that today's mathematics does not have the tools needed to tackle it.

Why, then, do we nevertheless believe in the conjecture? There is a heuristic, meaning to say informal, reason. It is that there are more "snakes" (by which we go down) than "ladders" (by which we go up). The ladders are longer, because when we go up we multiply by 3 and add 1, while when we descend we divide only by 2. But the number of descents is greater because every ascent is followed by a descent, while a descent is not necessarily followed by an ascent. An ascent happens when we come to an odd number, and after the multiplication by 3 (which keeps the number odd) and the addition of 1 we obtain an even number, after which the rules demand that we go down. An assumption that has no proof, but seems plausible, is that after every descent there still is a 50 percent chance to descend further, and after that, there is an additional 50 percent chance to descend, and so on. If this is so, it easily follows that for every ascent (that rises by a factor of about 3), there are, on average, 2 descents. Since these two descents mean going down by a factor of 4, every ascent times 3 is typically matched by a 4 times descent. Therefore, on average, we drop further than we rise, so that, eventually, there is a good probability of reaching 1. Of course this does not constitute a proof, because even if the probability of an event is low, it can still occur.

Another difficulty is that the numbers can go in a circle: there is no *a priori* reason why, say, if we begin with the number 537, the sequence will not eventually come around again to 537, just as when we begin with 1 we return to 1 (the sequence beginning with 1 is: 1, 4, 2, 1). To date no such circle has been found, other than the one beginning with 1, and with assumptions similar to those we mentioned, there is a good chance that no other circle exists.

This is a famous conjecture. Is it also important? At first glance, the answer is no. It isn't connected to any other mathematical topic, nor does it have any direct consequences. But, of course, this depends on the type of its solution, if it will ever appear. If the solution will show that this series is "random," in the sense described above — that a snake has the same chance to be followed by another snake as by a ladder, then we will understand something of value about the structure of numbers.

The twin primes conjecture

A prime number is a number greater than 1 that is not divisible by any number other than itself and 1. The first five prime numbers are: 2, 3, 5, 7, and 11. Prime numbers are important because they are the building blocks of numbers: every number can be written uniquely as the product of prime numbers (ignoring the order of the multiplicands). The components of a number can appear more than once, for example

$$1200 = 2 \times 2 \times 2 \times 2 \times 3 \times 5 \times 5 = 2^4 \times 3 \times 5^2$$

Ancient Greeks already knew that there are infinitely many primes. The kingdom of natural numbers owes its complexity to the infinite number of its fundamental building blocks.

Research in number theory was renewed in the seventeenth century, and special attention was drawn to the behavior of prime numbers. One of the most fascinating questions concerns their density: what portion of the numbers between 1 and a given number n do they constitute. For example, there are 25 prime numbers between 1 and 100, that is, one quarter of the numbers are primes. Between 1 and 1000 there are 168 primes, about one sixth. Between 1 and 1,000,000 there are 78,498 prime numbers, about 1 in 13.

These examples show that the higher the number, the smaller is the relative part of the primes among numbers below it. The primes become rarer and rarer. In the one thousand numbers between 1,000,001 and 1,001,000 there are fewer prime numbers than between 1 and 1,000. The question is, at what pace do they diminish? The above data shows that the rate of decline is not especially rapid. One million is a thousand times greater than one thousand, but in the jump between one thousand and a million, the portion of primes only dropped from $\frac{1}{6}$ to $\frac{1}{13}$, about two times.

We already mentioned Carl Friedrich Gauss in "To Discover or to Invent." In 1796, when he was nineteen years old, he proposed a conjecture regarding the density of prime numbers. It is that between 1 and any number n there are about $\frac{n}{\ln n}$ prime numbers, with $\ln n$ being the logarithm in the base $e = 2.718\ldots$ of n. (we will learn about the number e towards the end of the book. The concept "logarithm" is explained in the Glossary). This conjecture, one of the holy grails of the mathematics of the 19th century, was eventually solved independently in 1896 by two mathematicians, Jacques Hadamard and Louis de la Vallée-Poussin. The theorem that they proved became known as the "Prime Numbers Theorem." It states that not only are there infinitely many prime numbers in the world, but their portion among all the numbers is also not negligible. There are many prime numbers between 1 and n, for a large n.

There are other manifestations of the proliferation of prime numbers. Two famous conjectures point in this direction:

1. **The Twin Prime Number Conjecture**: There are infinitely many "twin" pairs of prime numbers, that is, pairs with a difference of 2.

   The first twin pairs of primes are:

$$(3,5),(5,7),(11,13),(17,19),(29,31),(41,43).$$

2. **Goldbach's Conjecture**: Every even number larger than 2 is the sum of two prime numbers.

The Twin Primes Conjecture states that the set of prime numbers is sufficiently rich to contain many prime numbers that are close together. According to Goldbach's Conjecture, the set of prime numbers is rich enough to express every even number as their sum. Both conjectures are very famous, and are usually bound together. The Goldbach Conjecture is the better-known of the two, possibly because it was named after a person. Christian Goldbach, a minor German mathematician, got into the pantheon of mathematics due to a letter he sent to Leonhard Euler that contained the conjecture. For a reason that will be explained in the next chapter, "Independent Events," no one doubts its veracity. Actually, typically the larger the even number, the greater the number of ways it can be expressed as the sum of two primes. The number 10 can be written in two ways as the sum of two prime numbers: $3 + 7$ and $5 + 5$. The number 100 can already be written as the sum of two primes in five ways: $3 + 97$, $11 + 89$, $17 + 83$, $29 + 71$, and $41 + 59$.

"Be wise, generalize," goes a famous mathematical dictum. Here is a generalization of the twin primes conjecture.

Conjecture: for every even number k there are infinitely many prime pairs whose difference is k.

In other words, there are infinitely many pairs of prime numbers with a difference of 4 between the pair; there are infinitely many pairs of prime numbers with a difference of 6; and so on. The number k must be even, because if the difference is odd, then one of the two numbers is even, and therefore it is not a prime number (unless it is 2). This conjecture is also similar in form to the Goldbach Conjecture, in which every even number is the *sum* of two primes (in the new conjecture, it is the difference between two primes). A major breakthrough was obtained in 2013, when the joint effort of many mathematicians culminated in a proof of the conjecture for all $k > 244$.

How can simple theorems have complicated proofs?

So, how can simple theorems demand complex proofs? This is a mystery, to which I can only attempt an explanation. I think that this is an optical illusion: the simple statements are those that draw our attention. Like adventurers who seek gold, and incidentally discover an entire continent, mathematicians, too, try to solve simply stated problems, and in the midst of doing so, discover complex theories. If they were to start from the end, from the theory, things would look differently. The simple statements in a mathematical theory are only a small part of the body of knowledge, but they are the most conspicuous because of their concise formulations. Our gaze is riveted to them, and it seems to us that they are the main focus. Actually, they only make up a small part of a big world, little pegs protruding out of the big rock.

But I must admit: every time I encounter a short theorem with a long proof, I am surprised anew.

Independent Events

Independence

There is good reason that mathematicians believe in the correctness of the Twin Primes Conjecture and in Goldbach's Conjecture. It is related to a concept that is important not only in mathematics, but in life in general: independence of events.

"Independence" means lack of causal link. Two events are independent if information on one adds no knowledge regarding the probability of the other. For example, many people believe that if they rolled a 6 in dice three straight times, then there is little chance that the next toss will also be 6. In reality, different tosses of the dice are independent: if the dice are fair, then the events: "first roll 6" and "second roll 6" are not causally linked. Of special importance is the realization that most events in the world do not depend on your desires. There is no connection between the results of a football game and your support of one of the teams. Blowing on the dice before the toss doesn't really help.

Here is another example: gender and eye color. Knowing that a person is blond increases the probability that his or her eyes are blue. Knowing that the person is a female provides no information regarding the color of her eyes. The percentage of blue-eyed people is exactly the same among women as in the entire population.

Independence and probability

Assume, for the sake of argument, that the proportion of people in the population (whether men or women) whose first name begins with A is $\frac{1}{20}$. That is, one out of every twenty people has a first name beginning with A. Problem: among all married couples, what is the proportion of couples in which the first names of both the husband and wife begin with A?

We can imagine two extreme cases. If all men whose first names begin with A were to take an oath not to marry a woman whose first name begins with A, there would be no couples like this. On the other extreme, if all men whose names begin with A were to marry only women whose names begin with A, then all the couples in which the husband's name begins with A would belong to this category, and they would constitute $\frac{1}{20}$ of all couples. A more realistic assumption, however, is that people do not choose their spouses based on the first letter of their name. So there is no connection between the two events: the first letter of the husband's name and the first letter of the wife's name are independent events. What proportion of the couples then have both names that begin with A? The husband's name begins with A in $\frac{1}{20}$ of the couples, and assuming independence of events, in $\frac{1}{20}$ of these couples the wife's name, too, begins with A. Therefore, the couples in which the names of both spouses begin with A constitute $\frac{1}{20}$ of $\frac{1}{20}$, which is $\frac{1}{400}$ of all couples.

This exemplifies a principle: when two events are independent, the probability of their joint occurrence is the product of their individual probabilities. Let us see this in the gender and eye color example. Assume that $\frac{1}{3}$ of the population has blue eyes, and that $\frac{1}{2}$ of the population are women. Assuming independence, the blue-eyed women are $\frac{1}{3}$ of the women, namely a $\frac{1}{3}$ of $\frac{1}{2}$ of the population, which is $\frac{1}{3} \times \frac{1}{2} = \frac{1}{6}$. Likewise, the probability of rolling two 6's in two dice is the product of the probabilities of rolling a 6 in each one, that is: $\frac{1}{6} \times \frac{1}{6} = \frac{1}{36}$.

Positive correlation, negative correlation

If two events are causally linked, then the occurrence of one changes the probability of the other. We then say that the events are "dependent." If the occurrence of the first increases the chances of the other occurring, we say that there is a positive correlation between them. For example, there is a positive correlation between being blond and being blue-eyed. If the occurrence of one event decreases the probability of the other, we say that they are negatively correlated. There is negative correlation between being blond and being brown-eyed.

If the correlation between two events is positive, the probability that both will occur is **greater** than the product of their individual probabilities. For example, the positive correlation between being blue-eyed and being blond means that if $\frac{1}{3}$ of the population are blonds and $\frac{1}{2}$ are blue-eyed, then more than $\frac{1}{6}$ of the population will be both blond and blue-eyed. As an extreme example, assume that all blonds have blue eyes. In such a case, the set of blue-eyed blonds is identical to the set of blonds, that is, it constitutes $\frac{1}{3}$ of the population, which is more than $\frac{1}{6}$.

Why is the Twin Primes Conjectures almost certainly correct?

The Twin Primes Conjecture states that there are infinitely many numbers n, such that both n and $n+2$ are prime. Why do mathematicians believe in it? To see this, let us explain why there are many twin primes between, say, 1 and 1,000,000.

The secret is that there is no obvious connection between the primality of a number n and the primality of $n + 2$. These two events should be independent. The fact that 101 is a prime number gives no reason to think that 103, too, is a prime number, or the opposite. As already mentioned, there are about 78,000 prime numbers between 1 and 1,000,000, which means that about 1 out of every 13 numbers in this range is prime. In other words, $\frac{1}{13}$ of the numbers up to a million are prime. If there is no dependence between a number n being prime and $n+2$ being prime, then for about $\frac{1}{13}$ of these prime ns, $n+2$ is also prime. Therefore, the portion of the numbers n up to a million for which both n and $n+2$ are primes is $\frac{1}{13}$ of $\frac{1}{13}$, which is $\frac{1}{13} \times \frac{1}{13} = \frac{1}{169}$. In other words, for about $\frac{1}{169}$ of the numbers between 1 and 1,000,000 (about 6,000 numbers), both the number and the number $n + 2$ are prime. So, assuming independence, there should be approximately 6,000 pairs of twin prime numbers between 1 and 1,000,000.

The Twin Prime Number Conjecture, with some spare

Actually, there are more. There are 8,169 such pairs. This means that our assumption of independence was not accurate. The dice are in fact stacked in favor of twins. The reason is that these two events are actually dependent, and *positively* correlated. A number n between 1 and 1,000,000 being prime means that there is greater chance that $n + 2$ is prime. The reason is that the prime numbers are not evenly distributed between 1 and 1,000,000. They are more concentrated among the small numbers. About $\frac{1}{6}$ of the numbers between 1 and 1,000 are prime numbers, while between 1 and 1,000,000 only about $\frac{1}{13}$ are, remember? Accordingly, if we know that n is a prime number, then there is a higher probability that it is small. Then $n + 2$, also, is small (remember, we are talking about numbers from 1 to 1,000,000, an addition of 2 is insignificant), and this increases the chances that $n + 2$ is prime. It follows from this that more than $\frac{1}{169}$ of the numbers n between 1 and 1,000,000 meet both conditions, that n and $n + 2$ are prime.

From million to infinity

Between 1 and 1,000,000 there are 8,169 twin prime pairs. A similar calculation shows that the range between 1 and 10,000,000 is likely to contain more than 50,000 twin prime pairs (the actual number is 58,980). This means that when going from 1,000,000 to 10,000,000 new pairs are added, namely, there are many twin primes located in the range between these two numbers. Similarly, there are twin-prime pairs between 10,000,000 and 100,000,000. New pairs appear at each step, which means there are infinitely many pairs.

Let me repeat: the assumption on which this argument is based, that there is no negative correlation between the primality of n and that of $n + 2$, is not rock solid. So, the argument does not constitute a proof. It only provides good reason to believe in the conjecture.

Part II: How Mathematicians and Poets Think

Tell all the truth but tell it slant.

Emily Dickinson

Poetic Image, Mathematical Image

Words

If there is anxiety in a man's mind, let him tell it.

Proverbs 12:25

Late in my professional life I made a partial career change, and began teaching mathematics in elementary schools. I came full of enthusiasm and new ideas. In particular, I believed in direct experience and addressing intuition. I thought that if the children would have hands-on experience, the abstractions would come by themselves. I quickly learned how wrong I was, and how essential is an interim stage that I missed: words. Human knowledge is built one layer on top of another, and words are the cement that binds them together. The first understanding of an idea is always intuitive, but in order to build the next floor one needs to stabilize this understanding by precise formulations. Man owes to words his ability to build skyscrapers of knowledge. Here is a little related secret: telling your trouble to others is helpful not only because of the release and the sympathy you may get. No less important is the fact that you clothed your problem in words. This helps not to tread in place. The next time you will start at the point you stopped last time.

But words are only the scaffolding for thought, not its engine. Purely verbal thought is barren. This is why abstract words cannot touch the roots of thought nor the profundities of emotions. To really reach people's minds and emotions, words must be used against themselves. This means, mainly, constructing images instead of abstractions, pictures that are close to the sources of thought. Mathematicians know this, as do poets. This is why images occupy such a central place in both domains.

The poetical picture

One generalization is worth a thousand examples.

Anonymous

One example is worth a thousand generalizations.

Anonymous

In his book *Twelve Conversations on Poetry* the poet Yaoz-Kest explains the difference between songs and poems: the song speaks of the general, the poem about the individual. The song may use abstractions, the poem is always concrete. Just like mathematics, it uses examples, that is, pictures, to deliver its message. Here is a famous example, Goethe's "Wanderer's Night Song II":

Over mountains yonder,
A stillness;
Scarce any breath, you wonder,
Touches
The tops of all the trees.
No forest birds now sing;
A moment, waiting —
Then take, you too, your ease.

"Wanderer's Night Song II," Johann Wolfgang von Goethe, trans. Christopher Middleton

Wandering in search of self discovery is not the invention of today's backpackers. In the romantic period, the wanderer who leaves civilization to seek his inner truth was eulogized in many poems. Wilhelm Muller's cycle of poems *Winterreise* (*A Winter Journey*), put to music by Schubert, tells the same story as in Goethe's poem. There too, the wanderer, a frustrated lover, longs to die — the meaning of the last two lines of "Wanderer's Night Song II." Goethe's poem, however, was the harbinger of the genre. Incidentally, Schubert also set this poem to music, twice.

The first six lines of the poem paint a seemingly pastoral scene: mountain peaks, treetops, birds. Actually, this is a sinister scene, everything being so quiet and still. The uneven meter and the changing pattern of rhymes amplify the sense of unease, reflecting the hero's disquiet. And then, in the last line, we understand: the threatening quiet is a reflection of the wanderer's desire to die, expressed as the wish to rest. The poem is thus conjured by four means. Two that relate to external form (rhyme and meter), and two of indirectness: the image, and the metaphor of death as rest.

Like the metaphor, the poetical picture is used for making an indirect statement. It projects emotions and thoughts on the external world.

David Fogel (1891–1944) was a master of images. Fogel was born in the Ukraine and, except for a single year in Palestine, spent his entire life in Europe. His poem "The Cities of My Youth," one of his last two, was written in 1941 in occupied France, in a state of sheer isolation and desperation, with the Nazi manhunt nearing. The mood, however, is not particular to this poem of his. All his poetry expresses a feeling of estrangement.

The cities of my youth,
Now I've forgotten them all,
And you in one of them.

In a rain puddle
Barefoot, you will yet dance for me
But you must have already died.

From my distant childhood
How I hastened to run,
Until I came to the white palace of old age —
And it is spacious and empty.

My fledgling steps
I shall never see,
Nor you, I shall not see,
Nor I of then.
The caravan of days,
From afar,
Moves on
From where to nowhere
Without me.

David Fogel, "The Cities of My Youth," in *Collected Poems*

This is a poem of loss and surrender, of the abandoning of desires and lusts that old age entails. But the detachment is not expressed abstractly. A prose author would write "I forgot my youth." Fogel writes tangibly, "I've forgotten the cities of my youth." Old age and death are both portrayed in a single picture, the white palace. And most moving is the last picture, of the caravan of days, that treads from one empty horizon to another, the poet being alienated even from this bleak emptiness: "Without me."

A mathematical picture: The number line

How do I think about my problems?
In a methodical and tangible way.

Karl Friedrich Gauss, mathematician

Mathematicians too, think in pictures. Words and formulas are used only afterwards, for communication and stabilization. Here is a small example. My daughter, who was seven at the time, was playing in the bathtub, and I came in and told her she had to finish her bath in three minutes. She asked for another five minutes. Okay, I negotiated, let's make it another four. She understood we were speaking about compromising in the middle and said, "In that case, I'll come out in 100 minutes." If you can find the middle between 3 and 100, I told her, you can stay for that amount of time. "51 and a half minutes" she said, without batting an eyelash (needless to say, she didn't stay that long.) To my question how she had calculated this, she answered, "Half of 100 is 50, and half of 3 is one and a half, so the middle between them is 50 and one and a half, which is 51 and a half." But she couldn't explain why she calculated the middle this way — that's how we do it, and that's that. Now, years later, I still don't know how she did it. But here is a picture that she may have had in her subconscious:

The middle between 3 and 100 is the middle between 0 and 103.

Our eye tells us that the middle between 3 and 100 is the middle between 0 (which is 3 units to the left of 3) and 103 (that is 3 units to the right of 100); and the middle between 0 and 103 is half of 103, which is half of 100 plus half of 3, just as my daughter had calculated.

The picture we used here is one of the most effective tools in mathematics, the "number line." This is a line along which numbers are marked at equal intervals. If we were to attribute it to a single person, it would be Nicholas Oresme (1323–1382), a French prelate and mathematician. Actually, the number line is a very natural idea, because it is the reverse of measurement. Measuring length quantifies geometry, while the number line does the opposite: it brings geometry to the aid of numbers. It gives tangible, geometric form to the concept of the number. Besides size, it also illustrates direction, the negative numbers being to the left of zero.

A well-known story about pictorial thought is told about the German chemist Friedrich August Kekule. Kekule struggled for a long time to understand the molecular structure of benzene (an oily, flammable hydrocarbon). He knew this molecule contained six carbon atoms and six hydrogen atoms, but he couldn't understand how they were linked. Until one night, he dreamt of a snake holding its tail, and when he awoke, he realized that the carbon atoms are arranged in a circle.

Another famous mathematical picture, the "Venn diagram," is used to illustrate sets. It shows the distribution of elements among the sets. The following Venn diagram shows three sets, A, B, and C. The shaded area, for example, represents the set of elements in C that is not part of A or B.

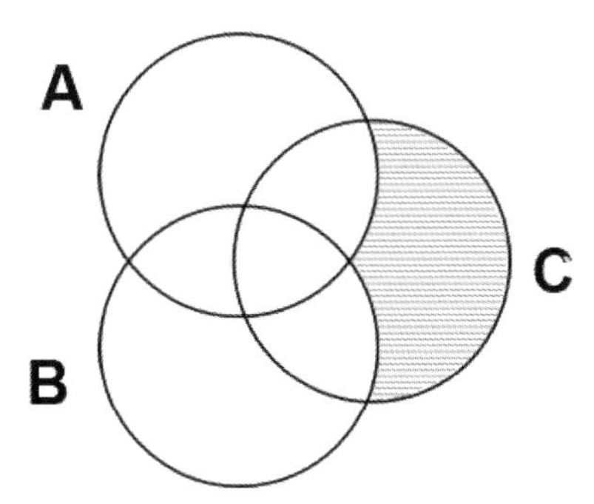

A coordinate system

The single picture that changed mathematical thought more than any other is the Cartesian coordinate system. "Cartesius" was the Latin name of René

Descartes (1596–1650), a French mathematician and philosopher, and the system is named after him, even though it was also discovered at about the same time by Pierre de Fermat (Curiously enough, simultaneous discoveries are not rare, which attests to what extent ideas float in the air at the time of their discovery). The Cartesian system is a two-dimensional extension of the number line, with two axes in the plane instead of one. While the number line represents numbers, the Cartesian system enables us to visualize pairs of numbers. The horizontal axis is usually called the "x axis," the vertical axis the "y axis," and each pair of numbers (x, y) corresponds to a point reached by moving from the origin, which is the intersection of the axes, x units to the right and y units up (if x is negative, the movement will be to the left, and if y is negative, it is downwards). For example, the pair (3, 2) corresponds to a point that is 3 units to the right of the origin of the axes, and 2 units up:

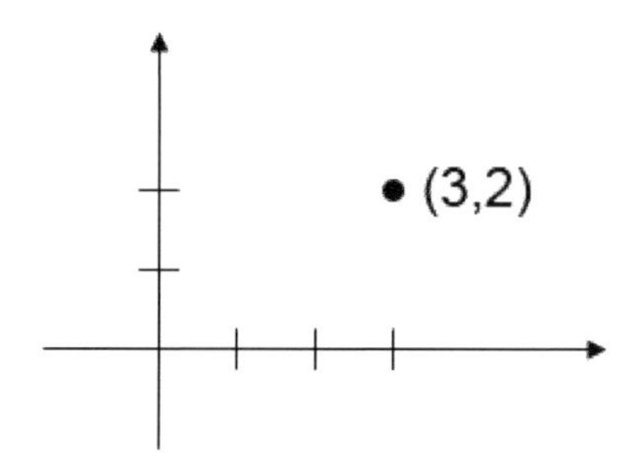

At first glance, there is nothing new here. Sailors used this system long before Descartes. To mark points on the globe they use two numbers, latitude and longitude. So what is so brilliant about this discovery? Descartes's innovation was not the discovery of the coordinate system, but in realizing its usefulness. Number pairs describe connections between numbers. As in life, numerical relations often link pairs (of numbers, in this case) rather than triplets or quadruplet. The coordinate system enables us to graphically depict these relations.

Take for example, the relation between a number and its square. The set of pairs that satisfies this relation is a collection of pairs of the form (x, x^2). For example, (0, 0) is such a pair, as is (3, 9), and also $(-3, 9)$, because $(-3)^2 = 9$. This also includes pairs of numbers that are not integers, such as (0.5, 0.25). If we draw all of these points on paper, the result is a "graph," which is a picture that depicts the relation. In this instance, the graph we obtain is a parabola. Each point on the parabola corresponds to one of the pairs.

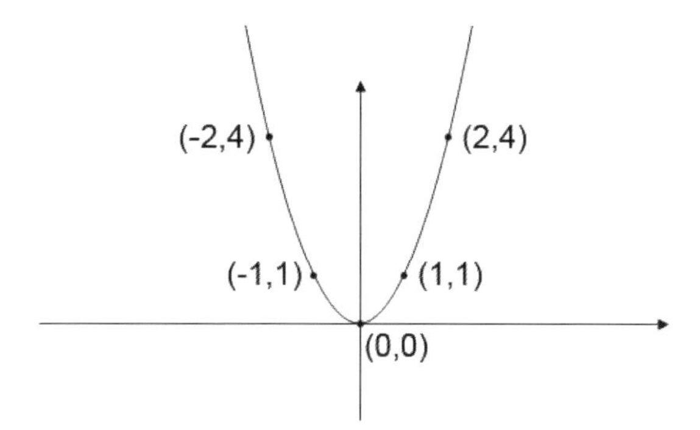

The relation between a number and its square is an example of a numerical function. A numerical function is a special type of relation between number pairs: a correspondence assigning a single number y to every number x. A real-life example of a numerical function: drive along the highway between Santa Fe, New Mexico and Denver, Colorado (the "Mile High City"). For every mile you drive, mark your elevation above sea level. The graph that describes this will show the elevation of the point as a function of the distance from Santa Fe.

Like every function, the "square" is an input-output machine. Given an input number, it outputs its square. The coordinate system helps us visualize the action of this machine. But it can also visualize relationships that are not functions. For example, "the distance of the point (x, y) from the origin, (0, 0), is 1" is a relation between x and y. There are pairs that satisfy it, and pairs that don't, just as among people there are pairs who are married and pairs that are not (and just as in the case of marriage, most pairs do not satisfy the relation). By Pythagoras' Theorem, the distance of the point (x, y) from (0, 0) is $\sqrt{x^2 + y^2}$. It follows that the points (x, y) at distance 1 from the origin are those satisfying the equality $\sqrt{x^2 + y^2} = 1$, which is true if and only if $x^2 + y^2 = 1$. Obviously, the points satisfying the condition lie on a circle with radius 1 and center at the origin. So, the circle describes the relationship given by the equation $x^2 + y^2 = 1$. The algebraic equality was transformed into a curve, and vice versa.

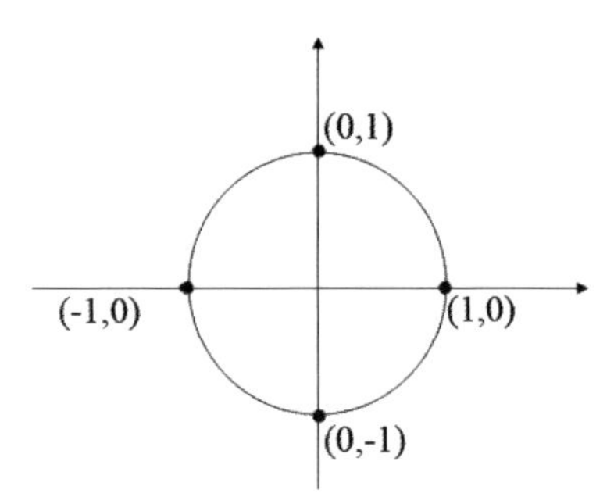

The circle describes a relation between the value x of a point and the value y of the point. The points on the circle are exactly the (x, y) points at a distance of 1 from the origin of the axes; by the Pythagorean Theorem, this is true only if $x^2 + y^2 = 1$.

Note that the equality $x^2 + y^2 = 1$ does not express y as a function of x. A function must assign precisely one value of y to each value of x, which is not true of this equality. For example, for $x = 0$, two values of y are suitable, 1 and (-1).

Thanks to Descartes, relations of this type become visible, and therefore easier to grasp. It is easier to visualize a circle than understand a formula. But there is yet another advantage to the Cartesian system, which was in fact Descartes's original aim. It goes in the opposite direction: geometric shapes are defined by formulas. For example, the collection of points (x, y) for which $x = y$ comprises a straight line, and the collection of points (x, y) for which $xy = 1$ comprises the two branches of a hyperbola.

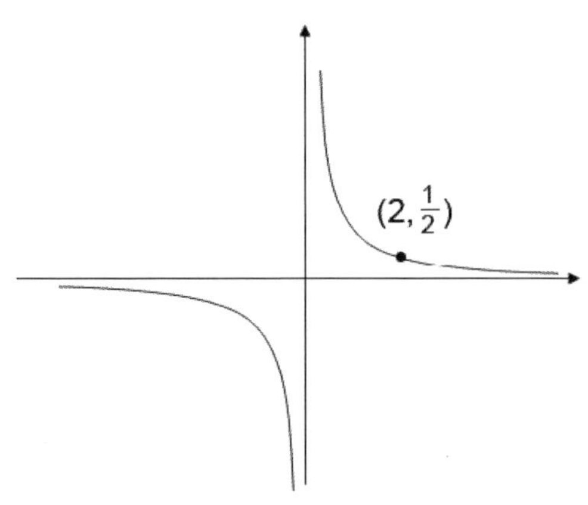

The graph describing the relationship $xy = 1$ between the variables x and y. It is called a "hyperbola."

Both algebra and geometry benefit from this system. Algebra gains the tangibility of geometric pictures, and geometry gains the use of algebraic tools.

René Descartes, French mathematician and philosopher (1596–1650). He chose a military career, but never took part in a battle. He spent his later years in Holland, until the queen of Sweden summoned him to be her private tutor. He did not withstand the harsh Swedish winter, and died of pneumonia shortly thereafter. His best-known contribution to mathematics is the Cartesian coordinate system, that is named after him.

The Power of the Oblique

In science one tries to tell people, in such a way as to be understood by everyone, something that no one ever knew before. In poetry, it's the exact opposite.

Paul Dirac, British mathematician and physicist

What did the poet mean?

To be a poem, a text should be oblique. Here, for example, is a poem by Rachel Bluwstein, "When It Comes":

And this is it? Only this?
Raising impatient eyes to this,
This, my lips, thirst to drink,
And with this, to warm the heart
in the chill of the night?
For this, to spurn God
and kick at His yoke?

This ... and no more ... no more

"When It Comes," Rachel Bluwstein

Those who know Rachel's life story (she is known to the Hebrew reader by her first name) can guess what this poem is about: a forbidden and unfulfilled love. But we don't have to know the details in order to grasp the mood the poem conveys, or in order to understand its message: "What is life, in comparison to a single storm of passion?" and, "There are things beyond the measure of everyday life." The poem's beauty lies in the sharp turn in the last line. Until this line, the poem is a protest of the poet against the world and against herself. And then, like in a punch line of a joke, everything is overturned. She comes to grips with her choice and with the

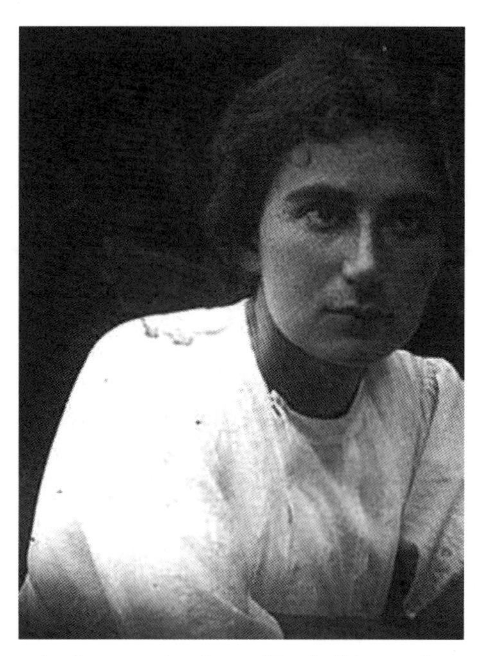

The poet Rachel Bluwstein (known simply as "Rachel"), was born in Russia in 1909 and immigrated to Palestine in 1909. She died in Tel Aviv in 1931, of tuberculosis she had contracted when she returned to Russia to treat the children of First World War refugees.

world, and accepts that the power of the heart is stronger than her. Due to the suddenness of the insight, and its minimalist statement ("no more... no more"), we perceive it like a feather's brush, and can pretend as if we really hadn't heard it.

Indirect proofs

Indirectness is not just an artistic means of the poem; it is an essential. A poem is never direct. We less expect to find indirectness in mathematics. But when it does appear there, it is always a source of beauty.

> A mathematician and his friend are walking in the
> forest. The friend boasts: "In a flash, I can know
> how many needles are on this pine tree."
> "How many?" the mathematician asks.
> "143, 547," says the friend, without batting an eyelash.
> The mathematician takes a handful of needles,
> and asks: "And how many now?"

This little story encapsulates several characteristics of mathematics. First of all, economy: the mathematician saves himself the trouble of counting the needles. He will subtract the new number his friend tells him from the first, and the difference should equal the number of needles in his hand, that can easily be counted. There is beauty in turning the boasting against itself. Making your opponent do the work for you saves energy, and hence is aesthetic. Economy of thought is a source of pleasure. Another source of beauty is the new element that comes out of the blue — subtraction. But probably most beautiful is the mathematician's indirect approach. He doesn't directly confront his friend's boasting. He doesn't count the needles, and at the end of the story this number is not known.

Some of the most beautiful arguments in mathematics are of this type. There is nothing more elegant than showing that something exists, without pointing at it concretely.

An irrational number to the power of a rational number

I already told you the story of the irrationality of $\sqrt{2}$. In the eighteenth century more difficult proofs were discovered for the irrationality of certain numbers. In 1737 Euler proved the irrationality of an important number called e (approximately 2.718; we will return to it later). In 1768 Lambert proved the irrationality of π, the ratio of the circumference of a circle to its diameter.

Then came more complicated questions: for example, is πe, the product of these two numbers, rational? There is no reason why it should be. As already mentioned, in a certain sense there are many more irrational than rational numbers (even though there is an infinite number of each), and therefore we could reasonably assume that πe is irrational. At our present state of knowledge, however, there no way of proving this. Is this important? Not particularly. The only reason why we care is because "it is there," or, as Hilbert put it: "We can know, [therefore] we must know." Questions of this sort are important only because the methods developed to solve them may turn out to be useful in other problems.

Hilbert was invited to the International Mathematics Congress held in 1900 to deliver one of the main speeches. In a daring move, especially considering his relatively young age (he was 38 at the time), he decided to give a lecture on the challenges facing mathematics during the coming century.

He chose 23 problems that he saw as central. His intuition proved correct. Most of these problems were solved during the twentieth century, and most were gateways to important developments. The seventh problem was proving the irrationality of certain numbers — for example e^π, or $2^{\sqrt{2}}$. This problem was solved in 1934 by the Russian mathematician Alexander Gelfond (1906–1968), and since then e^π is called the "Gelfond number." He also proved that $2^{\sqrt{2}}$ is irrational, but he had to share the glory with another mathematician, Theodor Schneider. It is called the "Gelfond-Schneider number." And what about π^e, for example? Is it, too, irrational? It is a safe bet that it is, but this is beyond our current state of knowledge.

These are deep problems, and I cannot tell you anything about the proofs. But here is a simpler question: prove that there are two irrational numbers, a and b, for which a^b is rational. Indeed, there are such numbers. And we shall show this without knowing the values a and b. Before beginning the proof, let me mention a simple rule of powers: $(x^y)^z = x^{yz}$. Here is a "proof" by example:

$$(x^2)^3 = x^2 \times x^2 \times x^2 = (x \times x) \times (x \times x) \times (x \times x) = x^6 = x^{2\times 3}$$

We want to prove the following:

Theorem: There are two irrational numbers, a and b, such that a^b is rational.

We have to distinguish between two possibilities. The first is that $\sqrt{2}^{\sqrt{2}}$ is rational. In this case, we are done: $\sqrt{2}$ is irrational, so we have an irrational number raised to an irrational power, producing a rational number. So, $a = b = \sqrt{2}$ satisfy the theorem.

The second possibility is that $\sqrt{2}^{\sqrt{2}}$ is irrational. Look at the number $(\sqrt{2}^{\sqrt{2}})^{\sqrt{2}}$. According to the rule of powers mentioned above, $(\sqrt{2}^{\sqrt{2}})^{\sqrt{2}} = \sqrt{2}^{\sqrt{2}\times\sqrt{2}} = \sqrt{2}^2 = 2$ (the last equality is nothing but the definition of the square root). Therefore, once again the irrational power $\sqrt{2}$ of an irrational number, $(\sqrt{2}^{\sqrt{2}})^{\sqrt{2}}$ (remember, in the present case we assume that this is an irrational number) is the rational number 2, which is precisely what we wanted.

In fact, we know which of the two possibilities is true. By Gelfond's Theorem, $\sqrt{2}^{\sqrt{2}}$ is irrational. The merit of out proof is that it didn't use this deep theorem.

The pigeonhole principle

Here is another mathematical theorem whose proof is not explicit:

In Santa Barbara there are two people with exactly the same number of hairs on their heads. (If you argue that this is too simple, because there are probably two completely bald people, then we can state a stronger proposition: there are two not completely bald people with the same number of hairs.)

The proof is based on what is known as the "pigeonhole principle": if 101 (or more) pigeons occupy 100 cells, then at least one cell will be taken by more than one pigeon. In general, if the number of pigeons is greater than the number of cells, two pigeons (at least) will have to crowd into a single cell. The general formulation of this theorem is: When more than n objects are divided into n types ("cells"), there will be at least two of the same type.

Can such a simple principle be of any value? The answer is "yes," if the cells are cleverly chosen. Returning to the hairs of the Santa Barbarians, it is known that a person has at most 100,000 hairs on his head. At the time of writing of this text, Santa Barbara has a population of around 104,000. So, we may safely assume that there are at least 100,001 people in the city who are not totally bald. Dividing the not-bald Santa Barbarians into types, by the number of hairs on their head (that is, the first type has a single hair, the second type has two hairs, and so on), we see that there are more people than types. Therefore, there will be (at least) two people of the same "type," that is, who have the same number of hairs.

Independent sets

A set of numbers is "independent" (a term coined only for our purposes here) if no number in it is divisible by another. For example, the set {3,5,6} is not independent, because 3 divides 6. The set {3,4,5} is independent, because 5 is not divisible by 3 or 4, and 4 is not divisible by 3.

> Question: How large can an independent set of numbers between 1 and 100 be?

As usual, a good advice is to begin with simple examples. In this case, replace 100 by a smaller number. The smaller the better, so ask first what the maximal size of an independent set of numbers between 1 and 1 is. Obviously, {1} is an independent set, and so the maximal independent set contains a single element. What about the numbers between 1 and 2? The set $\{1, 2\}$ is not independent (2 is divisible by 1), and so we can take only $\{1\}$ or $\{2\}$ — in this case, as well, the maximal size of the independent set is 1. Between 1 and 3? The set $\{2, 3\}$ is independent, and contains two elements. Between 1 and 4: the set $\{3, 4\}$ is independent, as is $\{2, 3\}$, and there is no independent set with 3 of these four numbers; so in this case, the answer is 2. Now let us skip to 10: the set $\{6, 7, 8, 9, 10\}$ is independent, as are $\{5, 6, 7, 8, 9\}$ and $\{4, 5, 6, 7, 9\}$, each of which has 5 elements. A simple check shows that there is no independent set of size larger than 5.

By these examples, if n is even, then the maximal size of an independent set of numbers between 1 and n is half of n. If n is odd, then the maximal size is half of $n + 1$. For example, it is easy to find an independent set of 50 numbers between 1 and 100: $\{51, 52, 53, \ldots, 99, 100\}$ or $\{49, 50, 51, \ldots, 98, 99\}$. And indeed, there is no independent set of size larger than 50. Here is an elegant argument showing this. We shall prove that a set with 51 elements is necessarily not independent. Formally:

> A set of 51 numbers between 1 and 100 necessarily contains two numbers, such that the smaller of them divides the larger.

The proof makes use of the pigeonhole principle. As always, the trick is in defining the cells. Between 1 and 100 there are 50 odd numbers, and for each of them we will define a "cell," namely a set of numbers.
The first cell consists of 1, 2, 4, 8, 16, 32, 64 — all powers of 2 below 100.
The second cell is 3, 6, 12, 24, 48, 96 — all multiples of 3 by powers of 2 (again, below 100).
The third cell is 5, 10, 20, 40, 80 — the multiples of 5 by powers of 2.
The fourth cell is 7, 14, 28, 56 — the multiples of 7 by powers of 2.

The cell corresponding to a given odd number will consist of the odd number times all powers of 2 (namely, times 1, 2, 4, 8, 16, ...). For example, the cell for the number 3 will include the numbers $3\times 1 = 3, 3\times 2 = 6, 3\times 4 = 12, 3 \times 8 = 24, 3 \times 16 = 48$, and $3 \times 32 = 96$. (We aren't going beyond 100.) The cell for 25 contains the numbers 25, 50, and 100; and the cell for 49 contains only two numbers: 49 itself, and 98 (multiplying by 4 already takes us beyond 100.) So our cells look like this:
1, 2, 4, 8, 16, 32, 64 (the multiple of 1 by powers of 2)
3, 6, 12, 24, 48, 96 (the multiples of 3 by powers of 2)
5, 10, 20, 40, 80 (the multiples of 5 by powers of 2)
7, 14, 28, 56 (the multiples of 7 by powers of 2)...

There are 50 odd numbers up to 100, so there are 50 cells. And every number between 1 and 100 appears in one of them. As an example, which cell contains 92? Divide 92 by 2, and we have 46, which is even, so we can divide it again by 2 and get 23, which is odd. So, $92 = 23 \times 2^2$, and therefore 92 appears in the cell of 23.

Recall what is our aim: we want to show that among any 51 numbers no larger than 100 there is one that divides another. The 51 numbers go into 50 cells, and by the pigeonhole principle, two of them belong to the same cell. But if two numbers belong to the same cell, the larger number will be divisible by the smaller. For example, the numbers 12 and 96 belong to the same cell, because $12 = 3 \times 2^2 = 3 \times 4$ and $96 = 3 \times 2^5 = 3 \times 32$. Since 4 divides 32, also 12 divides 96.

Co-prime numbers

The great Hungarian mathematician Paul Erdős was told about a child prodigy, Lajos Pósa, who already knew higher mathematics at the age of twelve. Erdős invited the young Pósa to a restaurant and asked him: "Prove that, in any set of 51 numbers between 1 and 100, there are two co-prime numbers." (Two numbers are called "co-prime" if they do not have a common divisor, apart from 1. For example, 5 and 9 are not divisible by any number larger than 1, so they are co-prime, while 9 and 12 have 3 as a common divisor, and therefore are not co-prime.) Pósa raised his head from his soup bowl and said: "In a set of 51 numbers between 1 and 100 there are two consecutive numbers." Naturally, two consecutive numbers are co-prime. If, for example, the smaller number is divisible by 3, the next one (the number $+1$) is not.

Here, too, the pigeonhole principle is at work. The simplest way of proving that there are two consecutive numbers among 51 numbers between 1 and 100 is to divide the numbers into 50 cells of consecutive pairs: $\{1, 2\}$, $\{3, 4\}$, $\{5, 6\}, \ldots, \{99, 100\}$. Of the 51 numbers in this set, 2 will have to belong to the same cell, meaning that they are consecutive.

Pósa left mathematical research at a young age. Erdős used to refer to mathematicians that abandoned research as "dead." But Pósa is very much alive: he devoted his life to the cultivation of gifted children, and produced generations of bright mathematicians.

Compression

One merit of poetry few will deny:
it says more and in fewer words than prose.

Voltaire, French author and philosopher, 1694–1778

One merit of mathematics few will deny: it says
more in fewer words than any other science.

David Eugene Smith

I have no time for a short letter,
so I am writing you a long one.

Blaise Pascal, French mathematician and philosopher, 1623–1662

The German word for poetry is *Dichtung*, meaning "compression." A short poem can contain a whole world. The American poet Ezra Pound said that "Great literature is simply language charged with meaning to the utmost possible degree." Compression is one of the poem's magician tricks. When many ideas are presented together, we do not follow all that is happening. Things happen too fast to be consciously registered.

Any poem can serve as an example. The choice I made, "Smells," is tribute to a forgotten poet. Noah Stern was born in 1912 in Lithuania, studied in the United States, and immigrated to Palestine in the 1940s. He published very little — his only collection of poems appeared after his death. He led a tormented life, served five years in jail for a murder attempt, and eventually committed suicide in 1960.

The lilac that grows in secret
The lilac that grows silently blue somewhere
Reminded me of illusions on one continent.
And disappointments on another.

But the heavy smells of oranges
Already come to pleasure and to torture,
Already come to give and to choke, as witnesses
Of life in this homeland.

"Smells," from *Among the Clouds*, Noah Stern

The Russian novelist Vladimir Nabokov said that smells are cloth hangers for memories. The two smells evoke worlds of emotions. A lot is compressed into the contrasts between the delicate, stealthy smell of the lilac and its dim blue color, and the heavy smell of the oranges and the blinding light of the southern country. The oranges give, like a mother's breasts, but also choke. Another expression of the poet's ambivalence towards his homeland is in the paradoxical wording "this homeland," as if there may be more than one homeland.

Compression is the secret of all art. It is compression that enables us to return again and again to the same work of art, finding something new at each visit. We are never tired of it, because so much is happening that we never understand it in full. The haiku poet Matsuo Basho (1644–1694) claimed that "a good haiku poem reveals only part of itself. We will never grow tired of a poem that reveals only half of itself." It should come as no surprise that lengthy books have been written on poems of only a few lines, or that thousands of words have been written on each note written by Beethoven.

But isn't this a case of "the eye of the beholder"? Does a poem or sonata really contain so much, or are their interpreters simply being inventive? A poem or sonata might be written very quickly. Even Beethoven, who was known for his countless drafts, wrote his sonatas in much less time than has been devoted to their analysis. Did so much really occur in his mind? The answer is a resounding yes — not only because he was Beethoven, but because our minds work much faster than we imagine. A short dream is capable of containing an entire world; every thought is the result of a complex process.

Why don't people understand mathematics?

I have been giving this lecture to first-year classes for over
twenty-five years. You'd think they begin to understand it by now.

John Littlewood, English mathematician, 1885–1977

We still have not touched upon the best-known, and most deterring, quality shared by mathematics and poetry: their difficulty. Both poetry and mathematics are hard to understand. The reason for students' difficulties is almost always the same: the teacher doesn't say all that he knows. He skips things. Even if he is aware of everything that came before, he doesn't have the time to spell them all out.

Conveying a lot of information in a single statement is what compression is all about. And it is this type of compression that is responsible for the difficulty in understanding poetry and mathematics. But there is a significant difference between the two: the compression in mathematics is vertical, while poetical compression is horizontal. In other words, in mathematics many stages, built like floors one upon the other, are hidden within a single statement. In poetry, many distinct ideas, not necessarily hierarchically ordered, are compressed into one expression. This is why the vague understanding of poetry causes no harm, while a hazy comprehension of mathematics gets back at us in a later stage, when the next floor is built.

Mathematical Ping-Pong

"How to solve it"

Gauss is like a fox who effaces his
tracks in the sand with his tail.

Niels Abel, Norwegian mathematician, 1802–1829

Why is it so hard to understand mathematics?
Because mathematicians think in examples,
but tell you the abstractions.

Anonymous

How do mathematicians think? Unfortunately, or perhaps fortunately, there is no recipe for this. A well-known book by George Polya, *How to Solve It*, describes thought strategies for solving mathematical problems. Although the book is replete with telling insights, reading it does not guarantee success in problem solving. The way to learn problem solving is not to read how others solved problems, but to solve them yourself.

Even though there is no magic formula, a basic trait of mathematical thought can nevertheless be put in words: it is conducted like a ping-pong game between examples and generalizations, between the tangible and the abstract. From examples we build generalizations, that, in the next phase, are confronted with other examples, which in turn lead to more accurate generalizations. This ping-pong game is not symmetrical, since shots in one direction — the abstract — involve magic, while the shots in the opposite direction — the examples — are more down-to-earth. There is no recipe for the generalization step. This is where we need illumination, the sudden discovery of hidden order in the world. In other words, this is where the beauty of mathematics is revealed.

Perhaps not surprisingly, the exact same can be said for poetry. There, too, a continuous ping-pong game is conducted between the concrete and the abstract. But there is a basic difference: in poetry the game is held within the poem itself. The abstract and the concrete coexist in the same lines. In mathematics, in contrast, we see only the results: the game is already over. It was there only at the stage of the struggle with the solution. The last shot in the solution is toward the abstract, and this is all that we, the spectators, see. The reader is like someone who is late for a play, arriving only for the last scene, after most of the characters have already made their exit. This is Abel's complaint against Gauss in the first quotation, and this is the anonymous student's complaint in the other.

So, in order to watch the ping-pong game of mathematical thought, we have to seize the moment. As an example, let me tell you how a sixteen year old discovered the formula for the sum of a geometric series. In the next chapter, we will give it a short and elegant solution. The high school student's method was a bit less elegant, but it is a fine example of the phases of mathematical inquiry, and provides a glimpse into the way mathematicians think.

The sum of a geometrical series

There are two common types of sequences: arithmetic sequences, in which each term is larger (or smaller) than its predecessor **by** a fixed number; and geometric sequences, in which each term is larger/smaller than the preceding number **times** a fixed number, called the "quotient" of the sequence. The origin of this name is that each term is the geometric mean of its two neighbors, meaning that its square is equal to the product of its neighbors. For example, in the sequence 2, 4, 8, 16, 32, 64, that has quotient 2 the term, 4 is situated between 2 and 8, and $4^2 = 2 \times 8$. In the chapter "Invention or Discovery" we learnt a formula for the sum of an arithmetic sequence (when we sum the elements we actually call it a "series"). There is also a formula for the sum of a geometric series, and I tried to lead the student to discover this formula.

I showed him the sequence 2, 4, 8, 16, 32, 64, ... , 1024, and asked him what was its rule, which he quickly found: each term is twice as large as its predecessor. I asked him if he could calculate the sum of the series, that is, 2+4+8+16+32+64+128+256+512+1024. The student was familiar with the formula for the sum of an arithmetic series, and tried the idea that worked there: finding the arithmetic mean of the terms (in an arithmetic sequence, the trick is that the mean is in the middle between the first and last terms).

He understood right away that this would not work here. In contrast with an arithmetic sequence, the middle between the first and last terms (between 2 and 1024) is not the arithmetic average of the terms of the sequence, since, for example, it is not equal to the middle between the second and the next to last terms (between 4 and 512). Another idea is needed.

A conceptual leap

I expected the student to try small examples. I thought that he would calculate the sums $2+4$, $2+4+8$, and so on, and find the regularity. But he surprised me with a true insight, made of the material from which mathematical discoveries are forged. "Let us look at the sum in the reverse order," he suggested. Like this: $1024+512+256+128+64+32+16+8+4+2$. "This is about 2048," he said (as usually happens at the beginning of a solution, things were still hazy in his mind). That is, about twice the first term (1024). Why? Because the addition of every term in the series halves the distance of the sum from 2048. We start with a sum of 0 (that is, our shopping cart was empty), and its distance from 2048 is, simply, 2048; the first term (1024) is half of this, and so after its addition, the distance to 2048 will be 1024, which is half the previous distance; 512 is half of the distance between 1024 and 2048, and after its addition, the distance between it and 2048 is 512 — again, half the previous distance. I think that the student had the following picture in his mind:

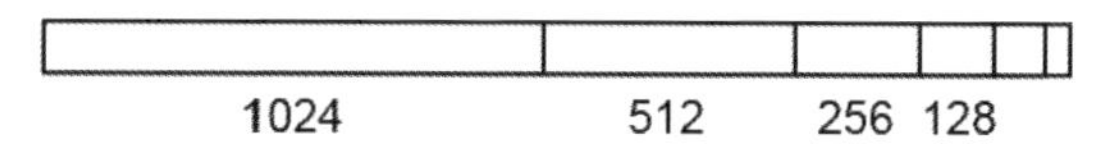

The length of the entire "rod" is 2048. Starting at the left, each step takes us half of the distance to the right end of the rod. In each phase, the distance from 2048 is the same as the size of the last step that we took.

At each step, the distance from 2048 is the same as the term that we just added. Now the student was already capable of calculating the exact sum. The distance of $1024 + 512 + 256 + 128 + 64 + 32 + 16 + 8 + 4 + 2$ from 2048 is the last term, that is, 2. Consequently, the sum equals $2048 - 2$, namely twice the first term in the series (1024), minus the last term (2).

The student found here the formula for the sum of a geometric series with a quotient of 2: the sum is twice the last term, minus the first term. In a formula, the sum is $2a_n - a_1$, with the terms of the series labeled as $a_1, a_2, a_3, a_n \dots$. This is the accepted notation for series: the first term is labeled as a_1, the second as a_2; and in general, the n-th term is denoted as a_n. I asked the student to check the formula himself, by examining the simplest example: what happens in a series with a single term? In this case, $n = 1$, and the entire series consists of a_1, which is both the first and last term. Using the formula that we found, the sum is $2a_1 - a_1 = a_1$, which is indeed the sum of the series consisting of a single element.

One more step

Next we moved on to a geometric series with a quotient of 3. For example, what is the sum of the sequence $10 + 30 + 90 + 270 + 810$, in which the quotient is 3, that is, each term is 3 times as large as the preceding one? The student tried his powers of generalization. If when the quotient is 2 the sum is $2a_n - a_1$, then when the quotient is 3 the sum should be $3a_n - a_1$, that is, three times the last term minus the first term. This is a smart guess, that even an experienced researcher would have made. I asked him to check an example. He wanted to take the example I had given him $(10 + 30 + 90 + 270 + 810)$, but I encouraged him to use a simpler example. He looked at a sequence with a single term — 10 (Remember the rule? There is no such thing as a too simple example). Its sum is, obviously, 10, while, according to the formula that he had guessed $(3a_n - a_1)$, the sum should have been $(3 \times 10) - 10 = 20$ (in this case, the last term, 10, is also the first). So, the formula he guessed does not work. The student wanted to abandon this line of thought, but I came to his aid here. It was probably not by chance that 2 appeared in the formula when 2 was the quotient. I encouraged him to try and compare the results of his guesswork formula $(3a_n - a_1)$ with the actual sums, in several examples. Take the sequence $10 + 30 + 90 + 270 + 810$, and sum up its initial terms:

The real sum	**The sum given by the false formula**
$10 + 30 = 40$	$3 \times 30 - 10 = 80$
$10 + 30 + 90 = 130$	$3 \times 90 - 10 = 260$

These sufficed for the student to discover the law: the formula that he had guessed gives twice as much as the truth. He must divide his formula by 2. And indeed, the formula for the sum is $\frac{3a_n - a_1}{2}$. A quick check with several examples showed the student that this indeed works.

Generalization

Now it is a short way to the formula for the sum of a geometric series with general quotient q. If, when the quotient is 3, the formula is $\frac{3a_n-a_1}{2}$, then when the quotient is q, the formula has to be $\frac{qa_n-a_1}{q-1}$ (presumably the 2 in the denominator of the formula $\frac{3a_n-a_1}{2}$ is $q-1$, putting $q=3$). For $q=2$ we get $\frac{2a_n-a_1}{2-1}$. But $2-1=1$, and division by 1 doesn't alter the number, so the formula gives $2a_n - a_1$, fitting the formula we found before. This explains the student's wrong guess: in the case $q = 2$ the denominator was hiding. It was hard to guess that there is a 1 in the denominator, which is actually $2-1$.

I then gave the student an additional example in which he could easily check the formula: $1+10+100+1000+10{,}000$. Here $q = 10$, and according to the formula that we discovered, the sum is $\frac{10\times 10000-1}{9} = 99{,}999 : 9 = 11{,}111$. Look at the sum and see why this is obvious.

Formal proof

A guess is not enough. We should prove it formally. At this point the student exhibited surprising mathematical maturity. For the proof, he told me, we need a formula for the terms of the sequence. This is not hard. If the first term is a_1, and the second is q times bigger, then the second term is a_1q. Similarly, the third term is the second term times q, namely, a_1q^2; and the fourth is a_1q^3. In general, the kth term is a_1q^{k-1}. If there are n terms in the sequence, then the last term is a_1q^{n-1}, and the sum of the terms in the sequence is $a_1+a_1q+a_1q^2+\cdots+a_1q^{n-1}$. Our guess is that this sum is $\frac{qa_n-a_1}{q-1}$. Since $a_n = a_1q^{n-1}$ this is in fact $\frac{qa_1q^{n-1}-a_1}{q-1}$, which is $\frac{a_1q^n-a_1}{q-1} = a_1\frac{q^n-1}{q-1}$. We therefore have to prove:

$$a_1 + a_1q + a_1q^2 + \cdots + a_1q^{n-1} = a_1\frac{q^n - 1}{q - 1}$$

Division of both sides of the equation by a_1 produces the equivalent formula:

$$(*) \quad 1 + q + q^2 + \cdots + q^{n-1} = \frac{q^n - 1}{q - 1}$$

Equality (*) can be proved by multiplying both sides by $q-1$. The left side will then be: $(q-1) \times (1+q+q^2+\cdots+q^{n-1})$, which equals

$$q \times (1+q+q^2+\cdots+q^{n-1}) - (1+q+q^2+\cdots+q^{n-1})$$

which, in turn, is $q+q^2+\cdots+q^{n-1}+q^n-(1+q+q^2+\cdots+q^{n-1})$. In this sum, almost everything cancels out, leaving only q^n-1. Note now that multiplying the right side of equation (*) by $q-1$ produces q^n-1, so indeed equality occurs in (*).

The Book in Heaven

Why are numbers beautiful? It is like asking why is Beethoven's Ninth Symphony beautiful.

Paul Erdős, mathematician

We have already met the Hungarian mathematician Paul Erdős (1913–1996). He became a legend in his own time, not only because he was a great mathematician, but also because of his unique lifestyle. He personified the image of the mathematician as detached from reality. He lived and breathed mathematics, and his interest in real life was limited and abstract. He traveled with only a light leather bag in his hand, relying on his hosts for all things material. Watching a movie, he incessantly bothered his neighbors: "What are they doing there [on the screen]?" He was very fond of writing letters, that usually began with sentences like "Let n be a natural number ... ."

Erdős' style of working, too, was unique. He would usually collaborate with his hosts on his many trips. Later in life, when he returned to Hungary after many years of self-imposed exile, he would work with the pilgrims to his mathematical shrine in the flat allotted to him in Budapest by the Hungarian Academy. He had more research partners than any other mathematician in history. Erdős preferred to think about elementary problems, and his greatness was in his ability to understand the profundity concealed in apparently simple problems. One of his earlier big achievements was an elementary proof for the Prime Numbers Theorem. As mentioned in the chapter "Simple Conjectures, Complex Proofs," this theorem had already been proved some 60 years earlier by Jacques Hadamard and Louis de la Vallée-Poussin, employing advanced means. An elementary proof was for many years a holy grail for mathematicians. But instead of making him happy, the discovey caused Erdős much heartache. He proved the theorem in cooperation with a Norwegian mathematician, Atle Selberg, and as was his custom, he immediately told the entire world of it. This led Selberg to (erroneously) believe that

Erdős was trying to claim the credit for himself. Echoes of the quarrel that erupted between the two reverberate to this very day.

Erdős had his own private language, with a special vocabulary. He called women "bosses," and men "slaves"; children were "epsilons" (after the Greek letter used in differential calculus to mark small numbers). Whenever he met a small child, he would ask for his age, show him a coin trick, and then move on to higher spheres. As someone who was born into the First World War and witnessed the horrors of the Second World War, he called God the "Supreme Fascist."

A proof from the book for the sum of a geometric series

Erdős used to talk about "the book in heaven," which includes all the mathematical theorems, each with its most elegant proof. Being "from the Book" is the greatest compliment a mathematical proof can receive. A friend of mine, who discovered such a proof, made a wise remark following his discovery: proofs from the Book aren't always born as such. They rarely spring from the forehead of their inventors in their full beauty. They often require elaboration to become pearls.

In the last chapter I described the process experienced by a student who discovered the formula for the sum of a geometric sequence. This was a prolonged process, and the proof was not particularly elegant. Here is the proof, after polishing.

Recall that we have to calculate the sum $a_1 + a_1q + a_1q^2 + \cdots + a_1q^{n-1}$.

Call this sum S. The secret is that multiplying the sum by q gives almost the same sum. Multiplied by q, every term becomes the next term — except, of course, for the last term, that does not have a next term. So, qS is almost S. What is the difference? qS has an extra term a_1q^n, the last term after multiplication by q; and it lacks the first term, a_1.

So, $qS = S + a_1q^n - a_1$. This is an equation with S as unknown. To solve it, shuffle terms: $qS - S = a_1q^n - a_1$, or: $S(q-1) = a_1(q^n - 1)$, and dividing by $q - 1$ we get: $S = \frac{a_1(a^n-1)}{q-1}$. This is the desired formula.

Paul Erdős, Hungarian and citizen of the world (1913–1996).

Poetical Ping-Pong

In the pingpong of questions and answers
not a sound was heard
except:
ping ... pong ...

Yehuda Amichai, "The Visit of the Queen of Sheba,"
Two Hopes Distant, trans. by Chana Bloch and Stephen Mitchell

Poetry is the other domain in which the play between the abstract and the concrete is essential. Like mathematics, poetry is an ongoing dialogue between individual instances and generalizations, between the tangible and the abstract, the low and the high. Metaphor, for example, is such a game: from the individual to the general, and back. The poet thinks of something specific, say, his lover's eyes. In his excitement, he wishes to give this a more general dimension, and he thinks about the general characteristics of eyes: softness, or their shape. In the next step, he returns to something else that is worldly, which has similar qualities: "Your eyes are doves." Note that the last shot in this game is in the direction of the tangible. This is a general feature of the poetical ping-pong: the heart of the poem is given to the concrete, and it is in this direction that the poem goes. This is the diametric opposite of the ping-pong of mathematics, in which the last shot is always toward the abstract.

As in mathematics, in poetry, too, the shots and their returns are so fast that in order to follow them we must freeze the moment and look at the game in slow motion. As an example, let us take Amichai's poem from which the quotation at the beginning of this section was taken. The poem is based on the legend in which King Solomon and the Queen of Sheba pose mental challenges to each other. For Amichai, these are the hide-and-seek games of lovers, that substitute for the real thing. Towards the end of the poem, however, the metaphor dissolves. The disintegration of the symbolism mirrors the breakdown of the lovers when the time comes to depart.

All the word games
lay scattered out of their boxes.
Boxes were left gaping
after the game.

Sawdust of questions,
shells of cracked parables,
wooly packing materials from
crates of fragile riddles.

Heavy wrapping paper
of love and strategies.
Used solutions rustled
in the trash of thinking.

Long problems
were rolled up on spools,
magician's tricks were locked in their cages.
Chess horses were led back to the stable.

Yehuda Amichai, "The Visit of the Queen of Sheba,"
trans. by Chana Bloch and Stephen Mitchell

The depiction of the games as shells transmits a sense of missed opportunity — the two protagonists themselves wanted something beyond the shell, but they did not reach it.

Concealed ping-pong

One of the most powerful poems on the Holocaust was written by Dan Pagis, a survivor:

Here in this carload
I am Eve
with my son Abel
if you see my older boy
Cain son of Adam
Tell him that I ...

Dan Pagis, "Written in Pencil in the Sealed
Railway-Car," *Transformation*, trans. Stephen Mitchell

One source of the poem's force is obvious — the nonstatement at its end, that leaves the reader hanging in air, and compels him to return to the poem's beginning. But the real strength of the poem lies in something else, less evident: the play between the abstract and the concrete. In these six short lines there are at least three transitions in each direction.

The poem begins with the tangible. It tells of a particular woman, in a railroad car that is not all cars, but a special one. Even the poem's title attempts to convey this, by insisting on the detail of the writing implement. We are so drawn to the woman in the car that we tend to forget the metaphoric nature of her name. And yet, the poem is not only about the suffering of this specific woman, and the poet gives her a name that represents all women, with her son who represents all children.

From this generalization the poem returns to the concrete: "if you see my older boy" — a simple, down-to-earth expression, the way a real mother would talk about her son. These words allude to the unbelievable — Eve still relates to Cain as a son, and even a beloved son. Poetry, as we know, can bear unresolved contradictions.

At this juncture the concrete use of the word "son" is replaced by an abstract meaning, as part of the wording "son of Adam" (the Hebrew, *ben adam*, also means a human being, and especially, a decent human being). This is obviously ironic — the last thing anyone could say about Cain is that he was a decent human being. But alongside the abstract sense, the words "son of Adam" also have a concrete meaning. Pagis reminds us that the metaphoric "son of Adam" had a concrete source — there actually was a person who was Adam's son. In poetry research this maneuver is called "metaphor reification" or "concretization."

This, however, is not the end of the ping-pong game, since the words "son of Adam" have an additional meaning: "You are your father's son, not mine." Couples sometimes joke — "see what your son has been up to." But Eve says this in all seriousness, as an actual statement of fact — the last move in the ping-pong game.

Laws of Conservation

You better cut the pizza in four pieces because
I'm not hungry enough to eat six.

Yogi Berra, baseball player

A law of conservation states that something — quantity, size, or ratio — is preserved, even when other factors in the picture change. For example, if you move the chair on which you are sitting, its position will change, but not the relations between its parts, and it will remain a chair. Thanks to simple laws of conservation like this, we can relate to the world in constant terms. There are more abstract laws of conservation, such as the conservation of number: if you take 4 stones and arrange them in a row, and then rearrange them in a square, their number will not change. Even more abstract is the conservation of matter. In a famous experiment the Swiss psychologist Jean Piaget (1896–1980) transferred a liquid from a wide container to a narrow one. Naturally, the liquid was higher in the narrow vessel, and when small children were asked if the quantity of the liquid changed, they answered: yes, there is more now, even though the liquid was poured from one vessel to the other before their very eyes.

The best-known laws of conversation are those in physics: the conservation of mass, energy, momentum (the product mass $\times$ velocity), and angular momentum. Elegant solutions can be found for many problems in physics using these laws. A less-known fact is that laws of conservation are used in mathematics as well. The difference is that the conserved elements are less tangible, which makes the laws even more beautiful.

Conservation laws usually not regarded as such are the rules for expanding or reducing fractions. Take a cake, and divide it into two halves. Unlike what Yogi Berra thought, the total quantity of the cake does not change, and we still have one cake. This means that 1 (one cake) is equal to 2 halves, or in numerical formulation: $1 = \frac{2}{2}$. Similarly, if we take $\frac{2}{3}$ of a cake and divide each of the two thirds into 5 pieces, the quantity of cake will remain the same. But now each third has become 5 pieces, each of which is $\frac{1}{15}$th of the cake. In other words, the two thirds together consist of 10 fifteenths, and so we have $\frac{2}{3} = \frac{10}{15}$.

How to become rich by using a law of conservation

The American Sam Lloyd (1841–1911) was one of the most ingenious puzzle creators of all times. He composed chess puzzles and mathematics puzzles, was an amateur magician and a professional ventriloquist, and included ventriloquism in his magic acts. His son would "read" his thoughts, when it was Lloyd himself speaking through his son's mouth. In 1875 he composed (some say, borrowed from another source) his most famous puzzle, the "15 puzzle," that is still popular today. It consists of a square with 16 smaller squares, on 15 of which are pieces bearing the numbers 1 through 15, with one square remaining empty. A piece can be moved to the empty square if it adjoins the empty square, that is, if it is alongside, above, or below the empty square. The puzzle is usually played by sliding the squares around, and trying to arrange them in order by a sequence of legal moves.

In order to boost sales, Lloyd offered a prize of $1000 (which was a considerable sum in those days, but not enough to arouse suspicion) to anyone who could exchange the places of the 14 and the 15.

1	2	3	4
5	6	7	8
9	10	11	12
13	14	15	

The original configuration

Even Lloyd himself couldn't imagine the consequences of his challenge. The hysteria that it set off even exceeded that which was triggered by the Hungarian cube a century later. People left their jobs and walked around in the street with the game in their hands. In France, a law was enacted forbidding playing the game at work. Lloyd became rich, without risking a

The desired configuration

cent, because he knew that the task was impossible. In the nineteenth century news didn't spread as fast as they do in the internet age. For some reason, no reporter thought to interview mathematicians, and much time passed before people learned of the impossibility of meeting Lloyd's challenge.

Before I prove this, I'll announce a similar competition for the readers of this book, a scaled-down version of Lloyd's challenge. Start with the number 1, and at each move add or subtract the product of two consecutive numbers. Anyone who manages to arrive at the number 10 will receive a prize of $100.

Example:

> As a first move, we could add to 1 the number 6, which is the product of two consecutive numbers: 2×3. Now we have the number $1 + 6$, that is 7. Now we can add, say, 20, which is 4×5 (another product of two consecutive numbers), for a new sum of 27. We can now, for example, subtract 12, which is the product of two consecutive numbers (3×4), giving 15.

Could a better selection of moves reach 10? The answer is no, and my $100 prize money is safe. Winning this game is impossible because of a law of conservation: what is conserved here is the parity of the number. It always remains odd. This is because the product of two consecutive numbers is always even, since one of the two numbers is even. We started with 1, an odd number, and we add or subtract even numbers. When an even number is added to or subtracted from an odd number, the result is still odd. This is why we will not reach 10, that is even.

Something similar is at work in Lloyd's game. There, too, a number is associated with every state of the board that remains odd throughout the game; while the number linked to the state in Lloyd's challenge (the switching of the 14 and the 15) is even. Therefore, Lloyd's money remained safe. The difference between Lloyd's challenge and the one I posed earlier on is that Lloyd's number is more hidden. His number was the number of *order changes* of pairs of pieces (I'll immediately explain the meaning of this term). Let us move along the board in a certain order, as in the next drawing:

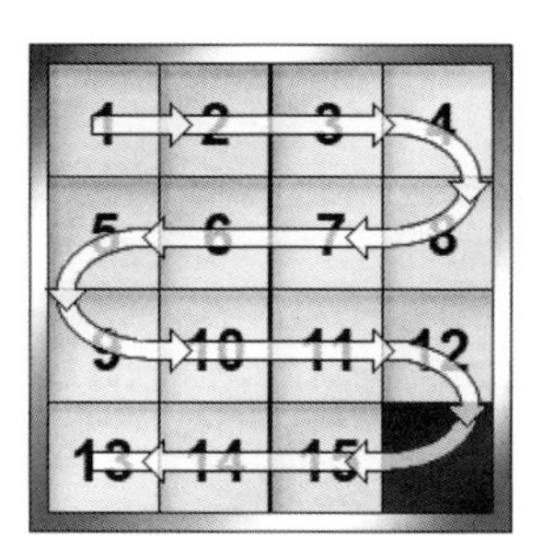

Even after the pieces have been moved, we will always go over them in the order dictated by the arrow. As can be seen from the drawing, when we go along the arrow we pass through the numbers in the following order: 1, 2, 3, 4, 8, 7, 6, 5, 9, 10, 11, 12, 15, 14, 13. An "order change" means a pair of numbers that is not in the right order. 3 and 10, for example, are in the right order in this sequence: 3 appears before 10, as in their regular order. But between 5 and 8, for example, there is an order change. In this sequence, 8 appears before 5, while the normal order calls for 8 to follow 5. In the wiggly sequence 15 appears before 14 — not the normal order between them, so here is another order change. How many order changes are there in this sequence in all? I'll list the number pairs in this sequence that are not in the right order — see if I've listed them all: (7, 8),(6, 8),(5, 8),(6, 7),(5, 7),(5, 6), (14, 15),(13, 15),(13, 14). There are 9 pairs, and so, in the original situation, there are 9 order changes of number pairs (as always, when following the arrow). In the desired situation, in which the 14 and the 15 change places, there is one less order change, since the (14, 15) pair is in the correct order (look at the arrow, and you'll understand why this is so). This leaves 8 order changes in the desired situation. The secret here is that the number of order changes remains odd, all the time, and therefore this situation — in which there are 8 order changes (an even number) — can never be attained.

Why does the number of order changes always remain odd? For the exact same reason that the number in the competition I suggested above remains odd: each move adds or subtracts an even number of order changes. Look for example at the following move:

This move does not add any order change, and does not remove any order change. For example, 12 appears before 13 in the wiggly order before this move, as it does after this move. And this is true for all other pairs. Simply, the order of the squares along the wiggly line hasn't changed.

Here is another example:

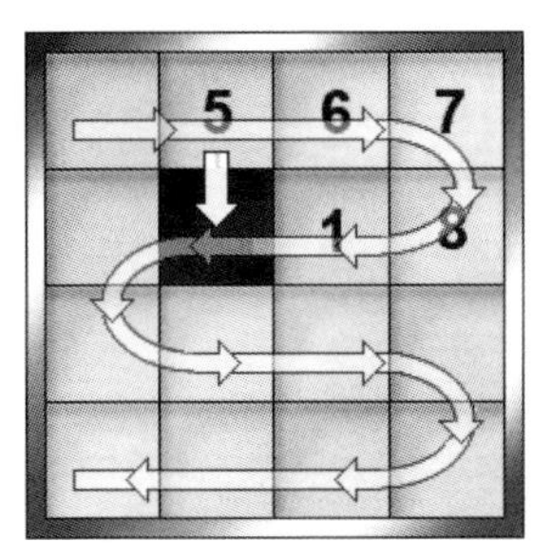

The move in the drawing alters the order changes of 5 only with the numbers 6, 7, 8, and 1. It adds 3 order changes (5 becomes out of order with 6, 7, and 8) and subtracts 1 order change (after the move, 5 is in order with 1, while before this move it was out of order with 1). So this move adds $3 - 1 = 2$ order changes. This is a typical example: in each move, the number of squares with which the moving square changes order along the wiggly line is 0, 2, 4 or 6. So, an even number of order changes will be added or subtracted. This means that if, in the starting position, the number of order changes was odd (to be precise, 9), then it will remain so throughout the game. Consequently, not only is it impossible to attain the state that Lloyd sought — any state with an even number of order changes is an impossibility.

Conservation of inner truth

In many Greek tragedies the hero tries to escape his fate, only to realize that it pursues him. The best-known tragedy in which this happens is *Oedipus Rex* by Sophocles. Oedipus, the young prince of the city of Corinth, hears a prophecy that he will kill his father and marry his mother. Horrified, he decides to flee the city. On his journeys, he meets a man at a crossroads and kills him in a fight. After this, he come to the city of Thebes, and at the gate of the city he learns of a monster, the Sphinx, with the head of a human and the body of a beast, who takes her toll on the city's inhabitants. The curse of the monster will not be lifted until the riddle she poses will be solved. Oedipus solves the riddle, and the grateful people of the city marry him to Jocasta, the widowed queen. Years later, when a plague rages unchecked in Thebes, and the oracle claims that Oedipus himself is responsible for this calamity, the truth is revealed to the king: he was a foundling, and his mother Jocasta had given him to a shepherd to raise, after she had heard the oracles utter the exact same prophecy that Oedipus had heard; and the person he had killed at the crossroads was his biological father.

Modern thought would view fate as symbolizing inner forces and wishes. Constantine Cavafy, the avowed Hellenist, conveys this message in many of his poems. The best known poem with this idea is probably "The City." It is a poem of "conservation of inner truth." External circumstances, it says, are not as important as who you really are.

You said, "I will go to another land,
I will go to another sea.
Another city will be found, better than this.
Every effort of mine is condemned by fate;
and my heart is — like a corpse — buried.
How long in this wasteland will my mind remain.
Wherever I turn my eyes, wherever I may like

I see the black ruins of my life here,
where I spent so many years, and ruined and wasted."

New lands you will not find,
you will not find other seas.
The city will follow you. You will roam the same
streets. And you will age in the same neighborhoods;
in these same houses you will grow gray.
Always you will arrive in this city.
To another land — do not hope —
there is no ship for you, there is no road.
As you scattered your days
you threw them away in the whole world, upon all seas.

Constantine Cavafy, "The City"

Cavafy lived from 1863 to 1933 in Alexandria, far from European culture centers. He was gay, worked as a civil servant, wrote in Greek, and gained his fame only after his death. But all these details tell us less than we can learn from this poem.

What makes this poem so beautiful? It is not the content, that is not new, but the way it is conveyed. Paradoxically, the message on the power of inner forces is expressed externally. Not "you will remain you," but "the streets will remain the same streets"; not "you cannot flee from yourself," but "the city will follow you"; not "it doesn't matter where you will be," but "in these same houses you will grow gray."

Constantine Cavafy (1863–1933), a Greek poet who lived in Alexandria.

An Idea from Somewhere Else

It can't be denied
Without symbols and simile there is no poem.

Nathan Alterman, "Footnote Poem," *Summer Festival*

The metaphor

Sometimes, when a description of a missing person is broadcast, I wonder why they bother. A description like "thin, short, athletic build, blue eyes, gray hair, thin nose, about sixty years old," is not really helpful. Except in cases of a peculiar characteristic, it is very difficult to identity a person on the basis of verbal description. If, however, the description were to state: "the missing person resembles President Bush Jr.," the chances of a positive identification will soar. A single key opens a door that remained closed to a pile of words.

What is the secret? In our minds we have a complete, ready-made picture of George W. Bush's appearance. Its reconstruction by breaking it into small details resembles moving a building brick by brick. If, instead, we move it en bloc, in all its detail, we can save much effort. Similarity to another person enables us to transmit a person's appearance without having to break it down into details. And this is not the only advantage: most of a person's facial features are too subtle to be transmitted in words. In other words, it is not always possible to even transfer the bricks. Some things can be conveyed only by analogy.

This is one of the two components of the best-known poetical device, metaphor. In Greek it means to transfer from another place. To describe a certain situation, we borrow the pattern of another, usually more familiar, situation. This is a powerful tool for transmitting information. So much so, that it is constantly used in everyday language. Our daily speech is saturated by metaphors, such as "it fell into my lap," or "a heart of gold." Even "saturated with metaphors" is itself a metaphor. See how much information is compressed into a single word: the feeling that, just as water saturates the soil, metaphors are everywhere; that just as the water is inseparable from the earth, so metaphors are so much absorbed in ordinary speech that we do not even notice them. Just as resemblance between people is so effective for description, a metaphor is capable of transmitting ideas that regular words cannot. Thomas Ernest Hulme, a British poetry researcher and poet (and a mathematician by education) who was killed in the First World War, claimed that "plain speech is essentially inaccurate. It is only by new metaphors, that is, by fancy, that it can be made precise." The features of the mold that is transmitted are sometimes too subtle to be transmitted in any other way. This explains the prevalence of metaphor in poetry. Here, for example, is a metaphor from "In the Twilight" by Hayyim Nahman Bialik, the Israeli national poet:

We were left with no friend or fellow
like two flowers in the desert.

Hayyim Nahman Bialik, "In Twilight"

An entire world is conveyed by a single brushstroke (here is another metaphor). The single line "like two flowers in the desert" communicates the sense of loneliness: the thirst for love; the lovers seeing no one except each other; their vulnerability; the contrast between their attitude to the world and their feelings for each other — and I certainly missed many other meanings. The desert is something familiar, and arouses in us a rich range of emotions and associations, and so a small tableau succeeds in conveying more than would many abstract adjectives. Here is poignant metaphor from the same poet:

One by one, with none to see, like stars towards dawn,
My secret desires vanished.

Hayyim Nahman Bialik, "One by One, with None to See"

Is there a more effective way of portraying how a person's childhood desires disappear during the course of his life, without his being able to point to the moment when this happened, than a comparison with the stars that disappear in the morning? The special beauty of this metaphor is also due to its revealing something about the image used. The unexpected awareness of the difficulty in discerning just when the stars vanish at daybreak is in itself beautiful. And the effect of the beauty of two truths revealed together is greater than the sum of the two separate effects.

But the metaphor would not have won its special standing if this was its only power. The true secret of its force lies in its being on both sides of the fence at the same time: it is an effective tool for transmitting information, and also for its concealment. It enables us to grasp matters without having to look them in the eye. It communicates the information innocently, as if this was about something entirely different. Some metaphors of everyday life have a similar role. We hear, for example, of "dropping out" of school, and we are not attentive to the poetical quality of this expression, that replaces blunt words like "leaving" or "expulsion."

Indirect expression has great force, no less for the writer than for the reader. Metaphor enables the writer to penetrate within himself and wrestle with questions which he could not confront head on. For example, poetry was Dan Pagis's way of contending with his childhood Holocaust memories that he did not otherwise dare to touch.

Once I read a story
about a grasshopper one day old,
a green adventurer who at dusk
was swallowed up by a bat.

Right after this the wise old owl
gave a short consolation speech:
Bats also have the right to make a living,
and there are many grasshoppers still left.

Right after this came
the end: an empty page.

Forty years now have gone by.
Still leaning above that empty page,
I do not have the strength
to close the book.

Dan Pagis, "The Story," *Synonyms*, trans. Stephen Mitchell

No reader of the last lines can leave the child-adult who faces the blank page, a metaphor for the adult confronting the emptiness of his lost childhood. The poem as a whole is built around the motif of entering and leaving a book. Look, for example, at the recurring line "Right after this came." At one point it refers to what occurs within the story (the owl delivers a speech), while in its next appearance, it speaks of what happens outside the narrative (the last, and blank, page follows). This is almost a play on words — identical words that have different meanings, in this case even of different types. Another play of being within and without the story is the green grasshopper (patently, green in two senses) who is a child in the story, while the most touching element of the poem is that the one who reads the story was a child at the time, but is no longer one, because his childhood was stolen.

An effective metaphor is always condensed, that is, there are many points of similarity between the tenor and the vehicle (the symbolized and the symbol). This density has an effect on both of the metaphor's roles: the transmission of information, and its concealment. On the one hand, the simultaneous transmission of many ideas means efficiency in communication; while on the other hand, when much information is delivered in a single effort, we are incapable of consciously absorbing it all, and most of its assimilation is subliminal.

In mathematics, just as in poetry, it often happens that ideas are brought from one realm to another. And just as an apt metaphor is finer, the more distant its tenor and vehicle are from each other, so too in mathematics: the solution is more elegant if the idea is brought from a more distant field. Number theory is known for the ideas that its draws from unexpected fields: geometry, complex numbers, differential calculus, and actually, from almost every other discipline of mathematics. In order, however, to present examples of this type, we must first learn of the division of mathematics into its various subdisciplines.

Three Types of Mathematics

Modern mathematics includes dozens of fields, and hundreds of subfields. Gauss, in the beginning of the nineteenth century, actually knew all the mathematics of his time. Hilbert, in the twentieth century, was familiar with all the mathematical fields of his time, even if he did not master them all. In the twenty-first century, a mathematician knows only a tiny fraction of the mathematics of his time. But even if mathematics has been divided into microscopic disciplines, its main branches have remained as they were. In this chapter I want to describe its basis division into branches.

Modern mathematics can be divided into three main branches: continuous mathematics, algebra, and discrete mathematics. The division is not exhaustive. It is difficult to exactly fit some fields, such as geometry or mathematical logic, into any of these categories. But it is a useful division, and basically a correct one.

Continuous mathematics

Continuous mathematics is concerned with things that change without leaps. Here is an example of a continuous question:

> A cat and a mouse are on a round field. These are a mathematical cat and mouse, that is, they are points. They cannot choose their speed — they both move at the same constant speed. On the other hand, they are free to choose the direction in which they move. Will the cat be able to catch the mouse?

The answer, that was proved by the Russian-English mathematician Abram Besikovitch (1891–1970), is typical of continuous mathematics. No, the cat cannot catch the mouse. But it can approach the mouse as closely as it wishes. In other words, if it can stretch out a paw — no matter how short the paw is — after enough time it will succeed in reaching the mouse. But if it doesn't have a paw to extend, it will only be able to approach the mouse, but not to catch it. The reason why the cat can come as close as it wants is that the mouse cannot run forever in the direction opposite to the cat, since it will run into the fence. The reason why, despite the mouse's inability to escape, the cat cannot catch the mouse is that when both are close to each other it is possible for the mouse to run almost directly away from the cat.

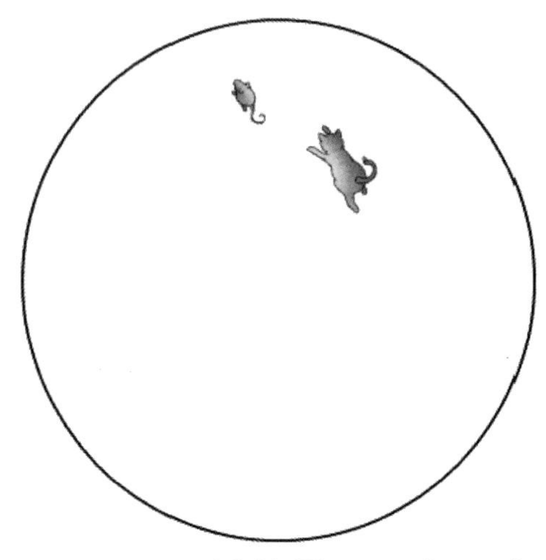

The cat and the mouse are on a round field. They can choose the direction in which to move, but not their speed — both move at the same speed. Will the cat be able to catch the mouse?

"As close as we want" is the essence of continuous mathematics. The central concept in this field is the limit, that means getting as close as you wish. For example, take the sequence of numbers 0, 0.9, 0.99, 0.999, ... None of the terms in the sequence equals 1, but they approach 1; therefore, 1 is the limit of the sequence. This means that for any measure of proximity (the length of the paw in the cat and mouse problem), beginning at a certain place the terms of the sequence are nearer to 1 than this measure. Beginning, for example, from the fourth term, the terms are closer than $\frac{1}{100}$ to 1. The limit of the sequence $1, \frac{1}{2}, \frac{1}{3}, \frac{1}{4} \ldots$ is 0, because, for any measure of closeness that we use, beginning from a certain term the numbers will be closer to 0 than that measure. For example, if the required measure of closeness is $\frac{1}{1000}$, the terms of the sequence are closer to 0 than this measure from the 1001th term.

Here is another example from continuous mathematics:

> A monk starts out from the bottom of a mountain at 8 a.m. He is making his way to the monastery, at the top of the mountain. He doesn't necessarily walk at a uniform pace, sometimes he walks faster and sometimes slower. He may stop every once in a while to

> enjoy the view. He reaches the monastery at nightfall, prays, and goes to sleep. At 8 o'clock the next morning he sets out on his way back, on the same path, and reaches the bottom at nightfall. Prove that there is a point on this path where the monk was at exactly the same time on both days.

Some people, when attempting to solve this problem, become bogged down in details: did the monk first walk slowly, and afterwards quickly? Or the opposite? They try to guess the location of this point. But all this is irrelevant. In order to solve the problem, we must change our perspective. Instead of thinking of a single monk on two days, we should think of **two monks on the same day**. If both set out at the same time and walk on the same path, one ascending and the other descending, they must meet; and "meet" means to be in the same place at the same time. This solution is based on a theorem that is very intuitive, but nevertheless needs proof. It is called the "mean value theorem," and may be formulated like this: someone who is riding in an elevator from the second shopping level of a mall to the -2 parking level must pass through a point that is on the ground floor. This is so, because the ride is a continuous process. If jumps were possible, that is, if our shopper could vanish at one point and reappear at another, then he could start at the shopping level and arrive at the parking level without ever actually being at the ground floor level. In a continuous world, this is impossible.

Algebra

Ask high school students what is this algebra with which they are spending so much time, and they will mumble something about "xs and ys." What are these, really? The answer is simple: they are names for numbers, and high school algebra is nothing more than calling numbers by names (or, more precisely, by letters). The need for letters arises in two contexts. One is when we speak about general numbers, that is, about any number. In order to relate to a general number, it must be given a name. Take, for example, the following rule: "the product of a number plus 1 multiplied by that same number minus 1 is the square of the number minus 1." Rather unwieldy, isn't it? Written in a formula, it is shorter and more understandable: $(x + 1) \times (x - 1) = x^2 - 1$. When a letter denotes a general number, it is called a "variable." The second instance in which we need to assign names to numbers is when the number is not known, and we attempt to find it by some given information. For example, "the number plus 1 is 3."

Using names, this can be written: $x + 1 = 3$. In this role, x is called an "unknown".

Classic algebra was developed by the Greeks, and later by the Indians. Based on their work, the Persian mathematician al-Khwarizmi wrote his book *al-Jabar*, which gave the field its name, and through which algebra was introduced to Europe. "Al Jabar" means balancing, and the book is called so after the technique of solving equations by operating the same way on both sides of the equation. For a lengthy period of time, algebraists devoted themselves mainly to solving equations. The first nonobvious type of equation to be solved was the quadratic equation, such as $x^2 - 3x + 2 = 0$. The ancient Babylonians and Egyptians knew of ways to solve this, as did the Chinese. The Greeks solved it geometrically (an example of this will appear in the chapter "Impossibility"). The solution of third degree equations, such as $x^3+4x^2-6x+1 = 0$, remained open, and was avidly sought by mathematicians for more than two thousand years. The problem was finally solved by several Italian mathematicians. The discovery of the solution was accompanied by one of the most tempestuous dramas in the history of mathematics. Two factors were responsible for this turmoil: the way in which the academic world operated at the time, and the personality of two of the protagonists, Niccolo Tartaglia (1500–1557) and Girolamo Cardano (1501–1576).

In the sixteenth century there were no academic journals, and communication of scientific ideas was done by letters. To the general public knowledge was disseminated in a somewhat strange fashion: public debates, which were a sort of academic wrestling matches. To win these competitions, many mathematicians would not publish their discoveries. And so it happened that the first mathematician to discover the solution of the third degree equations, Scipione del Ferro (1465–1526), kept his method secret. He wrote it for himself, and revealed it to his pupil Antonio Maria Fior only on his deathbed. Fior was a mediocre mathematician. He understood only part of the solution, namely, the solution for a certain type of equations, in which the quadratic term was not present, that is, there is no squared term (for example: $x^3 - 6x + 1 = 0$). He did not understand that not much separated this special case from the general solution. At that time Fior heard that Tartaglia, a poor teacher living in Venice, also, had found the solution to third degree equations. Certain of his superiority, Fior invited Tartaglia to a competition. Each of the contestants posed thirty problems to his opponent. Tartaglia found the method of solution of general third degree equations only during the competition, but once he did so, he solved all thirty of Fior's problems within two hours, and handily won the competition.

This is where Cardano entered the scene. Girolamo Cardano was a man of many talents. Besides being a leading mathematician, he was also a physician, the author of encyclopedias, an inventor (the transmission system he invented is in use to this day), and one of the first professional chess players in history. On top of all this, he was also a compulsive gambler, which led him to write the first book on mathematical probability — that also included a chapter on how to cheat. According to his own testimony, he was cantankerous. In 1570 he spent several months in prison on the charge of heresy: he was accused of casting Jesus' horoscope and attributing the events of his life to the influence of the stars.

Cardano attempted to find the solution to third order equations by himself, and when he failed he implored Tartaglia to tell him the secret. Tartaglia was himself a difficult man. At the age of twelve he almost died from a sword blow from a French soldier that slashed his face and mouth. This caused him to stutter ("tartaglia" in Italian means "stuttering"). He was suspicious, and refused for quite some time to divulge his secret, but eventually gave in, after a promise by Cardano to find him a patron. To ensure secrecy, Tartaglia sent the solution encoded in a poem. Cardano's assurance to attain support for Tartaglia was probably not sincere; at any rate, it never materialized.

Tartaglia quickly came to regret his giving in to temptation, and began a lengthy struggle, that was conducted in exchanges of letters. He was right: in the end, Cardano did not honor his promise of secrecy. About a decade later, Cardano heard that Tartaglia was not actually the first to discover the solution, having been preceded by del Ferro. Cardano no longer felt bound by his promise, and he publicized the discovery. Tartaglia was furious. He invited Cardano to a public debate, but the latter sent his talented pupil Lodovico Ferrari (1522–1565). Tartaglia lost the contest, and was consequently dismissed from the academic position he had recently won, after many years of poverty. In the end, he returned to his meager teaching position in Venice. At about the same time Ferrari also discovered the solution to fourth degree equations (such as $x^4 - x^3 + 4x^2 - 5x + 1 = 0$).

Polynomial equations continued to provide inspiration to algebra also in later years. But in the eighteenth and nineteenth centuries algebra made a sharp turn, and the meaning of the term changed. Modern algebra examines

operations. These are similar to arithmetic operations, but they can also be much more general, and relate to more abstract mathematical objects. An example is the movements in the plane. Among these movements there is an operation called "composition." Let us illustrate this. Draw a picture of something in the plane, say of a turtle, and select a fixed point of reference.

Now the turtle can be moved in different ways. These movements are called "transformations." It can be rotated around the point of reference, or moved in the plane — up, down, right, or left. Such a movement, that does not involve change of angle, is called "translation." Additionally, movements can be combined: performing one movement and then another. The composition of two transformations is dependent on their order. If the turtle is first translated and then rotated, it will arrive at a place different from the one it would reach if it were first rotated and then translated.

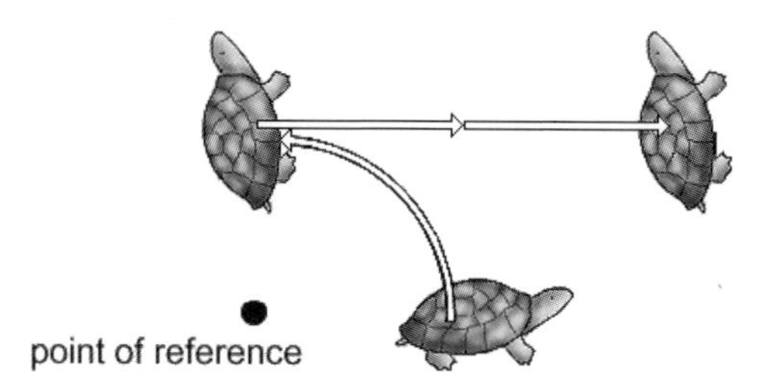

The composition of two transformations: rotation 90° counterclockwise, followed by translation to the right.

Motions in the plane are reversible. For example, if you move one meter north, you can go back by moving one meter south. Likewise, a 90° clockwise rotation can be reversed by a 90° counterclockwise rotation. The requirement of reversibility is the generalization of a well known property of operations with numbers: the addition of 7 can be reversed by subtracting 7; multiplication by 3 can be reversed by division by 3. A set (a collection of elements), together with reversible operations, is called a "group." The group is the

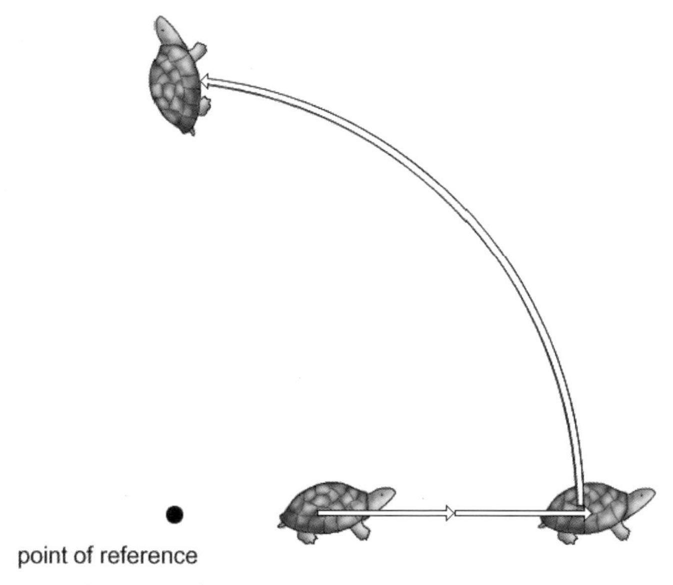

The composition of two transformations, in the opposite order: first translation to the right, followed by a 90° counterclockwise rotation. The result is different.

most basic algebraic object. This does not mean that groups are simple objects. They are surprisingly diverse and rich in structure.

Similarity of names usually indicates similarity in nature. What, then, is the connection between modern algebra, that examines abstract operations, and the study of equations? Why are they both called "algebra"? Not only because equations involve arithmetic operations, but also for a deeper reason. The French mathematician Joseph Louis Lagrange (1736–1813) discovered a surprising connection between the possibility or impossibility of solving a polynomial equation and operations on the set of solutions.

After the discovery of the solution to fourth order equations mathematicians tried desperately to solve fifth order equations. In the beginning of the nineteenth century a very surprising fact came to light: there is no general formula for solving fifth order equations. This was discovered by the Norwegian Niels Abel (1802–1829), who used Lagrange's ideas. Abel died of tuberculosis at an early age after a life of poverty and hardship, and left a mathematical inheritance that would "provide mathematicians with material for thought for two hundred years," as one mathematician of the period put it. Following him, the Frenchman Evariste Galois (1811–1832) showed just which equations can be solved and which not (not every fifth or higher degree equation is unsolvable; there are solutions for some equations of higher than fourth order). Galois's life, too, was tragic. He died in a duel at the age of 20. Recognition of his discoveries came more than a decade after his death. Galois used groups in his proof, and thereby further linked group theory to algebra.

Discrete mathematics

The third main branch of modern mathematics is **discrete mathematics**. This is the simplest of the three branches: its speaks only of sets and the elements belonging to them. There are no algebraic operations, and no concept of a limit or of getting infinitely close, the heart of continuous mathematics. Another name for discrete mathematics is "combinatorics," and until the beginning of the twentieth century, its common meaning was the counting of possibilities. For example: how many 3-member committees can be selected from among 10 people? In how many ways can 10 people be ordered in a row?

Until the middle of the twentieth century discrete mathematics was the stepdaughter of mathematics. It owes its present respectability to a single individual and to a technological revolution. The man who is accountable for the change was Paul Erdős, whom we have already met more than once. Erdős opened new directions in combinatorics, and showed the depths of which the field was capable. The technological revolution that brought combinatorics to the fore was the invention of the computer. For combinatorics proved to be the mathematics of the computer. A computer does not take continuous steps. In its memory it stores 0s and 1s, with no numbers in between. Going from one to the other is by jumps. As part of this field's growth, discrete mathematics came to include an increasing number of topics, and its ties with other mathematical fields were strengthened.

Topology

Mathematics is the queen of the sciences,
and arithmetic is the queen of mathematics.

Carl Friedrich Gauss

If a mathematics beauty contest were to be conducted, my guess is that arithmetic would take first place. This is the oldest mathematical field, and it deals with the most basic mathematical object — the number. There is a large gap between its depth and its seeming simplicity. It also has an especially large concentration of conjectures that even a child can understand yet their proof has eluded mathematicians for centuries. One of the most serious contenders for second place in the beauty contest is a more modern discipline: topology. The Greek word *topos* means "location," and topology is the "science of a location." A more precise definition is "the science of rubber sheets," since it examines the properties of sheets that are preserved under distortion. Topologically, a person is identical to a pretzel with several holes (the nostrils and the digestive system); and as far as topology is concerned, a person can take off his shirt without first removing his sweater. The difference between topology and geometry is that, in topology, it does not matter whether a line is straight or not, and distances between points are not measured. In other words, topology doesn't care if its rubber sheets are stretched and distances are enlarged. For it, the two sheets, the original one and the stretched sheet, are identical.

The fixed point theorem

A topologist is a geometer with his hands tied behind his back. He forbids himself to speak about distances. For him, the boundaries of a triangle and that of a circle are the same, since a triangle can be distorted to become a circle, or the other way around. But if distances aren't measured, then what remains to be said of shapes? One quality that topology examines is whether or not a shape has holes, and if so, how many? It develops tools to prove that two bodies are topologically equal, meaning that one can be transformed into the other by stretching, rotation, or reflection.

One of the best known topological theorems is the Fixed Point Theorem of the Dutchman Luitzen Brouwer (1881–1966). The theorem, dating to 1912, speaks of a ball in some dimension. A "ball" of radius R is the set of all points whose distance from a certain point (the center of the ball) is at most R. In one dimension this is a 2R long segment; in two dimensions, this is a disk (a "disk" is the inside of a circle, together with the boundary. By "circle" we mean just the boundary). In three dimensions, this is a regular ball, like the ones used in sports, and in four dimensions, it is a body that cannot be visualized. Brouwer's Theorem states that if we take such a body, distort it, move it, and stretch it — without tearing! — and if we leave it entirely within the same space that it previously occupied (that is, after the transformation no point is outside the place in space occupied by the body before the change), then there is a point that did not move. This is called a "fixed point," because it remains fixed in its place.

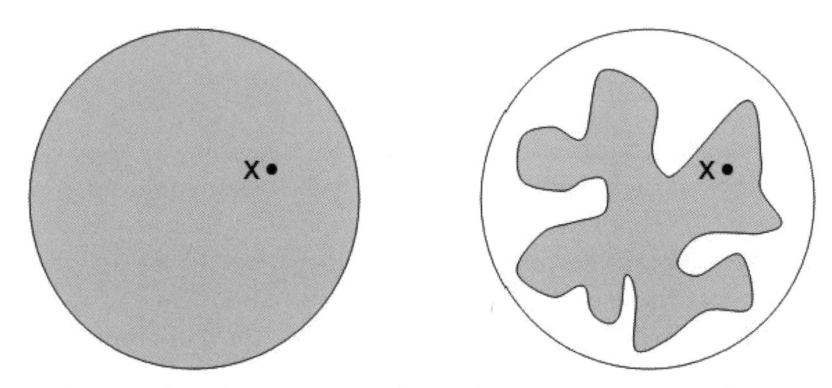

The gray shape on the right was obtained by the distortion and moving of the disk to the left (note that all points remained within the disk). Point x did not move — the distortion left it in place. Brouwer's Fixed Point Theorem states that in every distortion of a disk that leaves it within its original bounds, there is a point that stays put.

This theorem is no longer true if the body has a hole, for example if it is a ring. If a disk is rotated, its center remains fixed, but if a ring is rotated, there is no point that remains stationary — everything moves, since the center of rotation is not in the ring.

The two-dimensional case of Brouwer's Theorem can be demonstrated with sheets of paper (although paper can only be crumpled and moved, but not stretched, and therefore will not fully illustrate the theorem). For a topologist, a rectangular sheet is the same as a disk, since a disk can be distorted, without tearing, until it becomes a rectangle (just think about what can be done with a sheet of Play-Doh). Take two sheets of paper, of the same size and without holes, and place one exactly over the other. Now take the upper sheet, fold it, crush it, and rotate it as you wish — but without tearing it, and without any point on it going beyond the area of the lower sheet. According to Brouwer's Theorem, there is at least one point on the upper sheet that remains above the exact same point that it was above before. That is, it remained in place.

Its simple formulation, elegant proof (which I have not shown), and many applications, make Brouwer's Theorem beautiful. It is also a fascinating example of the application of one mathematical discipline to another. Although the theorem's formulation does not even hint at algebra, its simplest and most common proof is algebraic. The proof uses tools from a field called "algebraic topology," that was founded by the Frenchman Henri Poincarè (1854–1912). In the next chapter we will see a surprising use of the Brouwer's Theorem in discrete mathematics.

The Borsuk-Ulam Theorem

On the eve of the Second World War mathematics enjoyed a short-lived but dramatic blossoming in Poland. Mathematicians whose names are known to every present-day mathematician were active in the coffee houses of Lwow and Warsaw, the two major centers of mathematics research: Stefan Banach, Stefan Mazurkiewicz, Kazimierz Kuratowski, Alfred Tarski, Karol Borsuk, and many others. Topology in particular benefited from this spurt of activity. Stanislaw Ulam, later one of the fathers of the hydrogen bomb, was one of the younger of the group. He was not a topologist by profession, but he formulated a basic conjecture, which was quickly proved by Borsuk, and was named after them the "Borsuk-Ulam Theorem." First, an example of the theorem:

> At any given moment there are two antipodal points on the Equator with exactly the same temperature.

The Equator is just an example: we could have used any circle instead; and in place of temperature, we could have taken any other quantity, provided that it is continuous, that is, without leaps. Temperature does not "jump" — if at a certain point the temperature is, say, 10 degrees, then at neighboring points the temperature will be close to 10 degrees. That was an example of a one-dimensional case of the Borsuk-Ulam Theorem. Here is an example of the two-dimensional case:

> At any given moment, there are two antipodal points on the face of the earth with exactly the same temperature and the same humidity.

The face of the Earth is the boundary of a three dimensional ball. It is two dimensional, since every point can be specified using two numbers — the longitude and the latitude. In the two-dimensional case, the Borsuk-Ulam Theorem states that for every two continuous quantities there exist two antipodal points, such that both quantities are the same at the two points. Topologists speak also of higher dimensional balls, and when the number of dimensions rises, more quantities can be considered. For example: on the three dimensional face of a four dimensional ball (don't try to visualize this — it is a pure abstraction), for every three continuous quantities, there is a pair of antipodal points at which each of the quantities has the same value.

The proof of the theorem for two or more dimensions is difficult. The one-dimensional case, however, is quite simple. Remember, this case says that on the Equator (which is just an example of a circle) there are two antipodal points with the same temperature (which is an example of a continuous parameter or, in mathematical terminology, a "continuous function"). To prove this, we will draw a pointer with a head and tail, as in the following illustration:

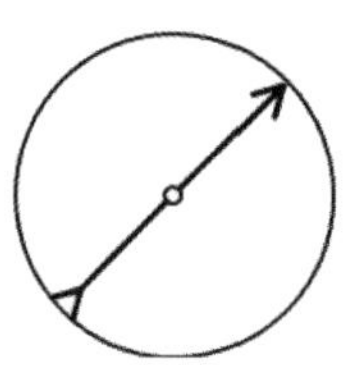

The Borsuk-Ulam Theorem states that if we rotate the pointer over the Equator, we will come to a situation in which the same temperature is measured at the pointer's head and tail. This is proved by measuring the difference between the temperature at the head and at the tail.

Place the pointer in any position, and calculate the difference between the temperature at its head and the temperature at its tail. If, at the position we chose, the difference is 0, that is, the temperature is the same at the head and the tail — then these are the two antipodal points with the same temperature, the existence of which we are after. We can therefore assume that this difference is not 0. Let us say, for example, that the temperature is 10 degrees at the head, and 3 at the tail, which gives us a difference of $10 - 3 = 7$ degrees. Now spin the pointer 180 degrees, to the position where the head and tail switch positions, all the while measuring the difference in temperature between the head and the tail. When the pointer arrives at its final position (that is, when the rotation of 180 degrees is completed), the temperature at the head (the former position of the tail) will be 3 degrees, and at the tail (where the head used to be), 10 degrees. The difference between the head and the tail is now $3 - 10 = -7$ degrees. And so, the difference changed from a positive value to a negative one. Since we assume that temperature is a continuous parameter, that is, without leaps, then according to the mean value theorem that we learned in the last chapter, at some point the difference has to be 0. But 0 difference between the head and the tail means exactly what we set out to prove — the existence of two antipodal points with the same temperature!

An application in discrete mathematics

A person's first response to the Borsuk-Ulam Theorem is liable to be — "So what?" As evidence of the theorem's importance, here is one of its many applications, in this specific case, in discrete mathematics. The problem I will describe makes no mention whatsoever of topology, but, surprisingly, the proof is topological.

The problem is called the "necklace splitting problem." Two thieves stole a necklace. The necklace is open (that is, it is not circular), and it contains beads of different types, with an even number of beads of each type. The thieves want to divide their loot fairly, with each thief receiving the same number of beads of each type. In order to do this they have to cut the necklace. But cutting the necklace will lessen its value, and so they want to make as few cuts as possible. How many cuts will they need? The Israeli mathematician Noga Alon used the Borsuk-Ulam Theorem to prove the following theorem:

The number of cuts needed does not exceed the number of bead types.

As usual — we should examine first the simplest example possible. In this case, a necklace with a single type of beads, in which case a single cut suffices — in the middle. The case of two bead types, too, is easy to prove.

Starting from 3 types, however, the only known proof uses the Borsuk-Ulam Theorem.

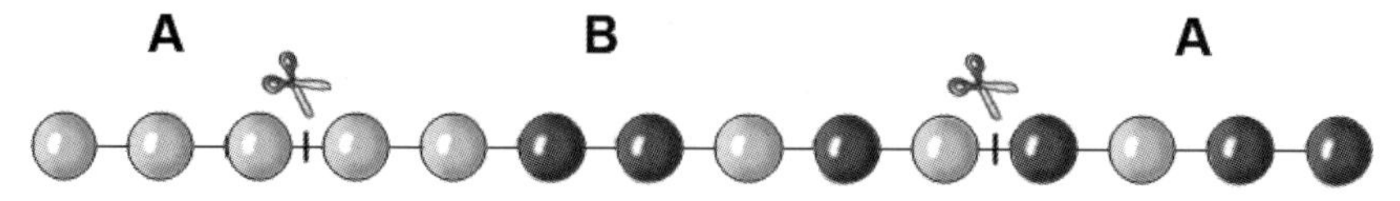

In this particular example, when the necklace is cut in two places (marked by the scissors in the drawing), the first thief receives the sections marked A, and the second thief receives the section marked B.

Matchmaking

Matchmaker, matchmaker, make me a match.

Fiddler on the Roof

1912, the year in which Brouwer proved his fixed point theorem, also marked a turning point in combinatorics. A cornerstone theorem of the field was proved then. Its discoverer, Ferdinand Frobenius (1849–1917), was both the blessing and the curse of the department of Mathematics at the University of Berlin. A blessing — because he was an outstanding mathematician; a curse — because he was contentious. The reluctance of other mathematicians to work with him was one of the reasons for the flourishing of Berlin's great rival, the University of Göttingen. Frobenius was an algebraist, not a combinatorialist (combinatorics hardly existed as a separate field at that time), and he worded his theorem in algebraic terms. When the Hungarian Dénés König proved a stronger version of this theorem a few years later and cast it in a combinatorial formulation, Frobenius ridiculed him for his "inferior terminology."

For many years König was the only one to recognize the importance of Frobenius' theorem. The theorem would become widely known only much later, in 1935, when it was discovered independently by the Englishman Philip Hall (1904–1982). The success of the new version (which was eventually named after Hall), might have been due to the intriguing name Hall gave it: the "Marriage Theorem." Sex sells, even in mathematics. Imagine, Hall said, a set of acquainted men and women: each man is acquainted with some of the women (possibly even with none). The men want to get married. The rule is that a man may marry only a woman with whom he is acquainted (these are mathematical men, and they don't have tall demands — the sole requirement is being acquainted). Naturally, the marriage must be monogamous, that is, a person (of either sex) may have only a single spouse. The question that Hall asked is: under what conditions can all the men be married? (Take note that we don't insist that all the women will get married — the problem was formulated before women's lib.) In order to understand the meaning of "under what conditions," look at the following example:

When two men are acquainted with only a single woman, they cannot be matched.

Acquaintanceship is represented by lines. There can be no marriage for all the men, since the two men are competing for a single woman, and so one man will have to forego getting married. This example points to a necessary condition for all the men to get married: the number of women must be at least as large as the number of men. Otherwise, following the pigeonhole principle, two men would be matched with one woman. This condition, however, is not sufficient for weddings for all to be held. For example, it might happen that there are no fewer women than men, but there is a man who is not acquainted with any woman — then this man cannot be matched (the right-hand case in the following drawing). Or, in a large group of men and women there may be two men who are acquainted with only a single woman (like the two left-hand men in the middle drawing), and then they cannot both marry. In each of the following drawings, A represents a set of men who cannot get married because they are not acquainted with enough women.

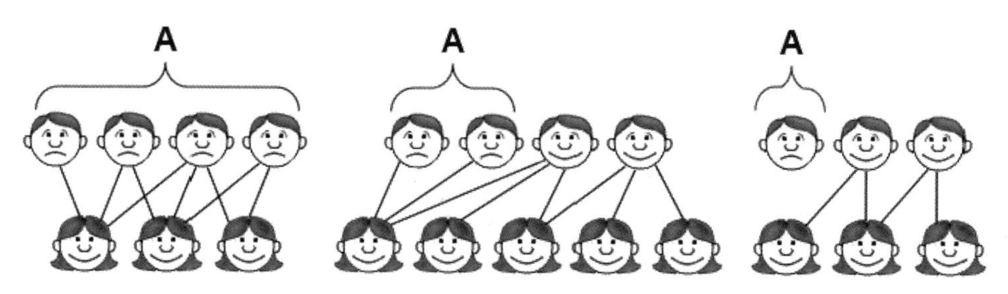

So, for a wedding of all the men to take place, it is necessary for every man to be acquainted with at least one woman, for every 2 men to be acquainted with at least 2 women, for every 3 men to be acquainted with 3 women, and so on. This is clearly a necessary condition. It is less clear that this is also a sufficient condition. Namely, if every k men are acquainted with at least k women, then all the men can get married. This is the content of Hall's Theorem:

> Hall's Theorem: There will be a wedding [i.e., all the men can be married] if and only if every set A of men is acquainted with a number of women at least the size of A.

Hall's Theorem is important because of its many uses, and because it served as the starting point for extensive and rich research in modern combinatorics. Theorems constructed in the pattern: "something (in this case, a wedding) exists only if every... (in this case, every set is acquainted with at least its size)" tend to be fruitful. They say "an object of a certain type exists, unless a clear and present obstacle to this exists." In our case: "The possibility of a marriage for each of the men exists, unless there is a set of men that is acquainted with too few women."

Ménages à Trois

Hall's theorem belongs to a rare species: theorems that are not hard to prove, and yet they possess depth and many applications. Experience tells that such theorems tend to have many different proofs. Pythagoras' theorem, for example, has hundreds of proofs, each with its own beauty. This is the case also with Hall's theorem. And surprisingly, the strongest proof, having the furthest reaching conclusions, is topological. The topological proof has applications beyond those of the combinatorial proofs.

One of these applications is to triple weddings. This is not what you think. I am not talking about two men and a woman, but about a man, a woman, and a third object, say their dog, or their lodging. Suppose, as an example, that we add to the men and women a third element — housing type. Instead of being acquainted with women, now every man will be acquainted with woman + housing (or woman + dog) pairs. A man's acquaintanceship with a woman + housing pair means that the man is willing to marry this woman, on condition that they will live in this housing.

Example:

> A man named Alan is acquainted with the pairs Alice + tent, and Betty + house. This means that Alan is willing to marry Alice on condition that they live in the tent, or with Betty on condition that they live in the house. He is not willing to live with Alice in the house, or with Betty in the tent. A second man, Bob, is acquainted with Betty + tent.

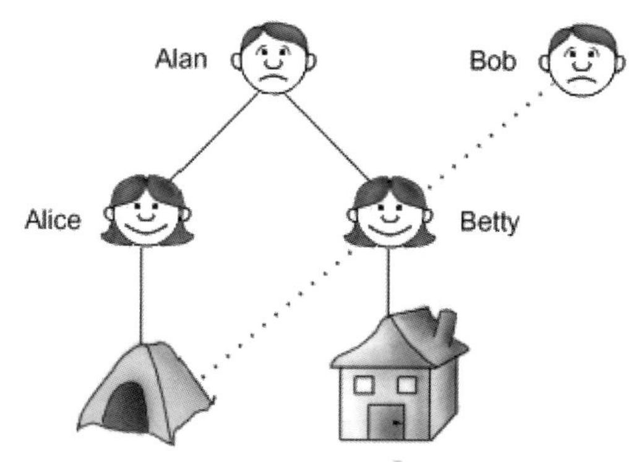

Alan is acquainted with the pairs Alice + tent and Betty + house, while Bob is acquainted with the pair Betty + tent. In this situation, there cannot be a marriage for both men.

A marriage in this case is the assigning to each man of a woman + housing pair that the man knows. Obviously, the marriage must be monogamous, which means here that two men cannot get married to the same woman, nor can two couples live together in the same housing. In our example, there is no such marriage. Bob can only get married to the one pair with which he is acquainted: Betty + tent. This means that if Bob gets married, Alan will remain single. He cannot be married to Alice + tent, because Bob took the tent for himself, nor to Betty + house, because Bob married Betty.

Let us now return to the general case, and assume for a moment that there is a marriage. In this case each man is acquainted with the woman + housing pair that he has chosen for his marriage. If so, then (say) every 3 men are acquainted with 3 such pairs, the ones that they have chosen for their marriages. These pairs are disjoint, meaning that all the women in them are different, as are all the housing types. Consequently, if a marriage of the men is possible, then every set of k men is acquainted with at least k

disjoint pairs. In the case of Hall's Theorem, in which the men married only women, this condition suffices. Does it also suffice in this case? That is:

> When, for every number k, every k men are acquainted with k woman + housing disjoint pairs, is a marriage guaranteed?

The answer is "No." In order to see this, all we have to do is look at the above example. Each man is acquainted with at least one woman + housing pair (Alan is actually acquainted with two disjoint pairs), and two men together are acquainted with two such disjoint pairs: the two together are acquainted with the two disjoint pairs Alice + tent and Betty + house (Alan by himself was already acquainted with these 2 pairs, and so Alan and Bob put together as a group certainly are acquainted with them). Nonetheless, there cannot be a marriage for both. Therefore, more is needed. And, here, "more" means double: if we double our requirement (in fact, a bit less than double), we will obtain a sufficient condition for the marriage.

> Theorem: If every k men are acquainted with $2k - 1$ disjoint woman + housing pairs, then there is a marriage for all the men.

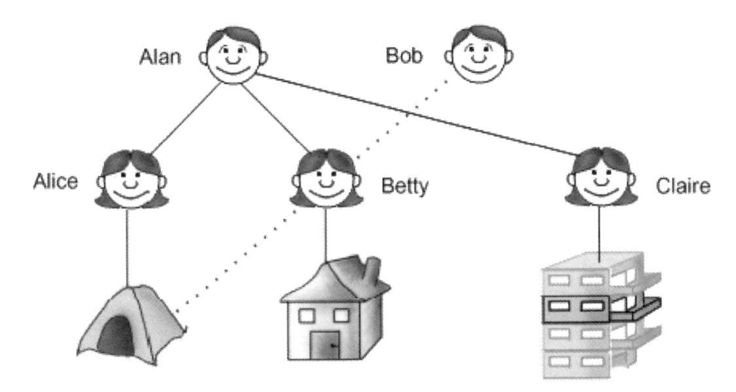

When 2 men are acquainted together with 3 disjoint pairs and each man is acquainted with at least one pair, the marriage is assured.

For example, if we add to this example one more pair (Claire + apartment), with which Alan is acquainted, then this requirement is met. For $k = 1$, the requirement means that every man must be acquainted with $2k - 1 = (2 \times 1) - 1 = 1$, namely, every man must be acquainted with at least one pair — this condition is met in the example. For $k = 2$, the condition means that 2 men together must be acquainted with $2k - 1 = 2 \times 2 - 1 = 3$ disjoint pairs, and this condition is also met: the two men are acquainted with the 3 disjoint pairs Alice + tent, Betty + house, and Claire + apartment. And indeed, both men can be married: we match Alan with the pair Claire + apartment, and Bob with the pair Betty + tent.

And here we are in for a surprise: the proof of this theorem is topological. Topology is not mentioned in the theorem, and is totally unexpected. Nonetheless, the only proof known uses topological tools — actually, a certain version of Brouwer's fixed point theorem.

Imagination

Fantasy and imagination

The gift of fantasy has meant more to me than
my talent for absorbing positive knowledge.

Albert Einstein

The English poet and essayist Samuel Tayler Coleridge claimed that poetry finds "similarity in difference." The poet William Wordsworth (1770–1850) used an almost identical definition in his philosophical poem "The Prelude": poetry discerns the similarities between things that look different to the passive observer. Imagination is the ability to find features shared by two seemingly distant objects. This ability is common to poets, mathematicians, and scientists.

In the early seventeenth century a poetic movement that its opponents called "Metaphysical poetry" emerged in England. It was characterized by sophisticated metaphor and the discovery of unexpected similarities between disparate objects. The leading figure in the movement was John Donne (1572–1631). Here is a famous passage from one of his poems, that bears the strange name "A Valediction Forbidding Mourning." The bond between the souls of the poet and his lover is compared to the relationship between the two arms of a compass. The analogy goes on and on, and every time that we think it has been exhausted, yet another point of similarity emerges.

If they be two, they are two so
As stiff twin compasses are two;
Thy soul, the fix'd foot, makes no show
To move, but doth, if th' other do.

And though it in the centre sit,
Yet, when the other far doth roam,

It leans, and hearkens after it,
And grows erect, as that comes home.

Such wilt thou be to me, who must,
Like th' other foot, obliquely run;
Thy firmness makes my circle just,
And makes me end where I begun.

John Donne, "A Valediction Forbidding Mourning"

The beauty of the last two lines lies in their multiple possible interpretations: does "makes me end where I begun" mean that the lover aids the poet to find his true inner voice? Or perhaps to return to his infancy?

Mathematical similarity

Poetry is the art of calling the
same thing by different names.

Anonymous

Yes, and mathematics is the art of calling
different things by the same name.

Jules-Henri Poincaré, French mathematician

One of the most famous reports by mathematicians of their creative process was related by Poincaré. One day, he was walking in the street, not thinking about anything in particular, at least not consciously, and he boarded a trolley car. The moment he put his foot on the trolley's steps, the light flashed on. He realized that the problem that had occupied him for weeks was identical to another problem, whose solution he knew.

Like in poetry, mathematical discoveries are often made by finding similarities. Here is a threefold example: three problems that, superficially, seem different, but have an identical underlying hidden structure. All three are well-known: the calculation of the area of a triangle, the sum of the numbers between 1 and a given number n, and how far an object accelerating at a constant pace will travel.

First, for the simplest of the three, calculating the area of a triangle. The area of a triangle is half the product of the side by its height. How can this

Jules-Henri Poincaré, French mathematician (1854–1912), the father of modern topology.

be proved? One way is by first realizing that it is enough to prove this for a right angle triangle. Every triangle can be divided into two right angle triangles, like this:

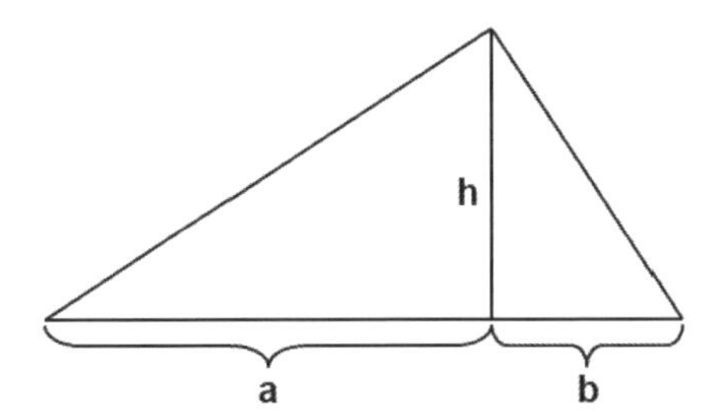

If we prove the formula for each of the two right angle triangles that were formed, then the area of the left-hand triangle is $\frac{1}{2}ah$, the area of the right-hand triangle is $\frac{1}{2}bh$, and the area of the entire triangle is therefore $\frac{1}{2}ah + \frac{1}{2}bh = \frac{1}{2}(a+b)h$. Since the length of the bottom side is $a+b$, this means that the area is half of the base multiplied by the height, as we wanted to demonstrate. In an obtuse triangle, the height line is outside the triangle:

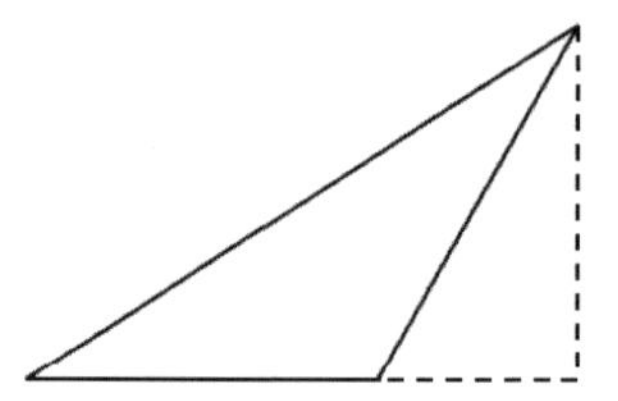

In this case, the area of the triangle is **the difference** of the areas of the right angle triangles.

So, all we have left to prove is the formula for a right angle triangle, like this:

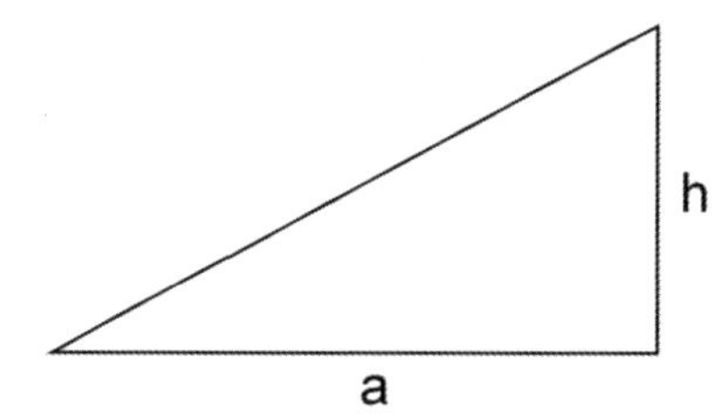

Here is one way of proving this. Note that the distance of a point on the upper side from the base increases at a constant rate as the point moves to the right. The point furthest to the left is at a height (distance) 0 from the base. If the entire upper side were at this height, it would fuse with the lower one, and the area of the line formed would be 0. The point furthest to the right is at a height of h above the base, and if the entire upper side were at this height, it would assume the shape of a rectangle, the area of which is the base times the height, that is, ah. The average height of the upper side is the middle between 0 and h, namely, $\frac{1}{2}h$. The area of the triangle is the product of the base times the average height, namely, $a \times \frac{1}{2}h$, which is $\frac{1}{2}ah$, which is what we wanted to prove.

In the chapter "To Discover or to Invent" we learned how Gauss calculated the sum $1+2+3+\cdots+n$. In order to show its likeness to the calculation of the area of a triangle, I will present the solution somewhat differently. Let us add 0 to the sequence, and write it as $0+1+2+3+\cdots+n$. Adding 0 doesn't change the sum. The first number is 0, and the last one is n. Since the numbers grow at a fixed pace, their average is the middle between 0 and n, that is, $\frac{1}{2}n$. After we added 0, the number of terms in the sequence is $n+1$ (before we added 0, there were n terms). The sum of the terms of the sequence is the number of terms multiplied by the average of the terms (this average is $\frac{1}{2}n$). That is, the sum is equal to $\frac{1}{2}(n+1)n$ — just like the calculation of the area of a triangle: "half of the length of the base times the height."

Gauss's method was a bit different. He matched 1 to n, 2 to $n-1$, 3 to $n-2$, and so on. The next drawing shows the calculation of the sum of the numbers from 1 to 6. 1 is joined to 6, 2 to 5, and so on.

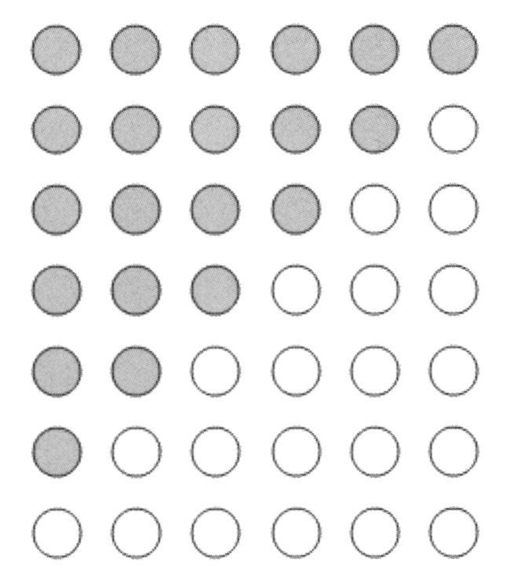

The sum of the numbers from 1 to 6 is the number of circles in the lower triangle. As we see from the drawing, this number is half the number of circles in the rectangle, that is, half of 6×7, namely half of 42, which is 21. This parallels the classic proof of the formula for the area of a triangle, as illustrated in the next drawing:

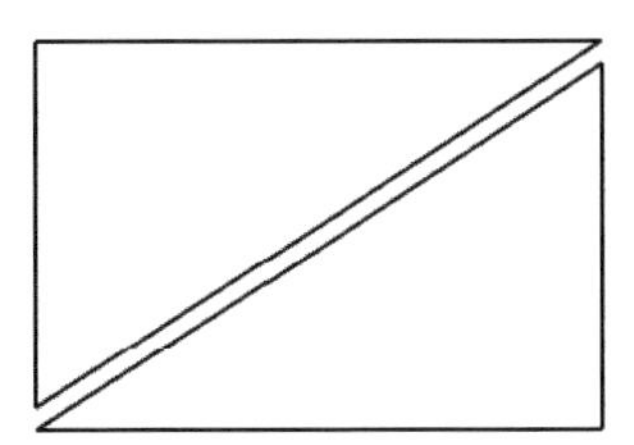

The two triangles, the upper and the lower, complement each other to form a shape of fixed height. This yields a rectangle, of height h with area ah. The area of the triangle is half this, that is, $\frac{1}{2}ah$ — which is exactly Gauss's proof!

Finally, a physics problem: the calculation of the distance traveled by an object accelerating at a constant pace. We will stipulate that the object begins at a speed of 0, moves for t seconds, and during this entire time its acceleration is a (that is, each second it increases its speed by a meters/second). How far does the object travel? We said that it begins at a speed of 0. If a meters/second are added to its speed each second, then at the end, after t seconds, its speed will be at meters/second. The average speed during its journey is the average between 0 and at, which is $\frac{1}{2}at$. The distance is the product of the time multiplied by the average speed, that is, $t \times \frac{1}{2}at = \frac{1}{2}at^2$, which is a formula known to anyone who studied physics in high school.

As we see, it is not only the formulas that are similar ($\frac{1}{2}$ appears in all of them), the ways of arriving at them, too, are analogous — they all use averages. I didn't realize this connection when I was in high school, and I understood it only when my son was in high school. Anyone who is aware of the similarity between the problems understands all three better: understanding is linking.

Isomorphism

Pushing the "Power" button on a radio changes its condition, from "off" to "on," or the opposite. It doesn't take any child long to learn that the same holds true for a computer, a television, or a game console. This is an example of "isomorphism." That is, structural likeness: two phenomena sharing the same hidden structure. The radio, the computer, and the television are "isomorphic" in terms of their on/off mechanism. We can add a mathematical example. If we take the numbers 1 and (-1), the operation of multiplication by (-1) reverses the on/off status, just like pressing an electrical switch: when 1 is multiplied by (-1), it reverses and becomes (-1), while (-1) multiplied by (-1) becomes 1.

In a party my daughter asked me what is the difference between the buffets at the two sides of the hall. Being a mathematician, I answered "they are isomorphic." She asked me what does this mean, and by way of an answer I suggested that we play the following game. Two players, each in turn, picks a number between 1 and 9, which had not been selected so far by either of the two players. Each player collects the numbers he chose, and the winner is the player who has three numbers that total 15. Here is an example of such a game. Call the players A and B. We record their moves, as well as the numbers they have collected so far:

A - 5 (the set of numbers collected so far is $\{5\}$)
B - 7 (B has collected the set $\{7\}$)
A - 6, his set of numbers chosen so far being $\{5, 6\}$. Now A threatens to play 4, which together with 5 and 6 will total 15. So B has no choice:
B - 4. Now B has $\{4, 7\}$. B poses no threat, because in order to get to 15 from $4 + 7$, 4 must be added, but 4 is already taken. So A is free to make the following move:
A - 1 Now A has at his disposition $\{5, 6, 1\}$. At this point, A poses two threats: to choose 8, which complements the 1 and the 6 he already has to 15; or 9, which, together with 5 and 1 (which he also possesses), total 15. B can counter only one of these threats. If, for example, B takes 8, A will choose 9 and win. If B chooses 9, A will select 8 and win. So A's win is guaranteed.

I played this game with my daughter. My son, who sat kibitzing, butted in: "You're playing tic-tac-toe!" He was right. Let us draw the familiar magic square, in which each row, column, and diagonal totals 15:

2	9	4
7	5	3
6	1	8

Choosing a number between 1 and 9 is actually choosing a square in this board. Instead of selecting a number, player A can draw an X in the corresponding square, while player B marks an O. Somewhat surprisingly, every threesome that totals 15 is a row, column, or diagonal in this square. There are only eight threesomes of different numbers from 1 to 9 that total 15, and all appear here. So a winning threesome in the game I proposed is just the same as winning at tic-tac-toe. Let us see this by illustrating the rounds of the above game between A and B on a tic-tac-toe board (on the boards in the illustration, going from left to right; the "X" starts).

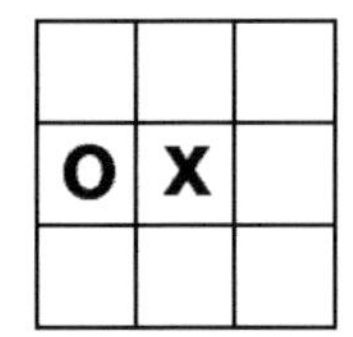

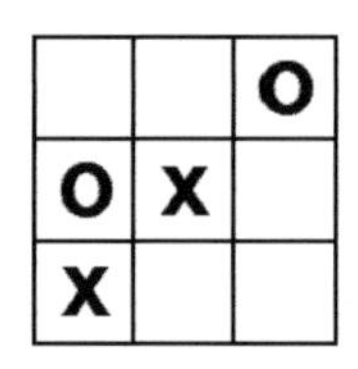

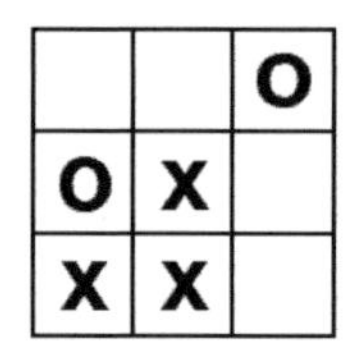

The first three steps taken by players A and B, from left to right. In the first round A chose 5, which is represented by the X in the left-hand drawing. B responded with 7, which is represented by the O in this board. After the third round, the "X" player has two possibilities of winning, only one of which can be blocked by the "O" player.

On the face of it, the number-choice game and tic-tac-toe seem different, but a proper abstraction shows the identity between the two: they are isomorphic. I asked my son how he discovered this, and he said that he identified two similar elements in both games: the threesomes, and the double threat. In tic-tac-toe, too, the way to victory lies either in the opponent's oversight or in a double threat, that is, two threats only one of which can be blocked.

"That reminds me" is one of the more common paths leading to mathematical discoveries.

A Magic Number

Strangely, of all the real numbers the Goddess of Mathematics picked two and assigned them a special role. One of them, π, which is about 3.141, was already well known in antiquity. This is the ratio between the circumference of a circle and its diameter, a ratio that already the ancients knew is the same for all circles. The symbol π was first used in 1706 in a book by William Jones, as the first letter of the word "perimeter." The number π naturally appears in geometric formulas, but, surprisingly enough, also in number theory. The significance of the second special number, marked as e, with a value close to 2.718, was discovered only in the seventeenth century. Its meaning could be comprehended only with tools from differential calculus, which developed then. Its importance quickly became apparent. The first to name this number (as b, rather than e) was the German mathematician and philosopher Gottfried Leibniz (1646–1716). It was named e by the Swiss Leonhard Euler (1707–1783), the leading mathematician of the eighteenth century. Contrary to the natural guess, Euler did not pick the first letter of his own name. He wanted to use a vowel, and since he had already used the letter a for some other value, he chose the second vowel in the alphabet.

It soon became clear that e is no less important than π. It appears in varied contexts, and in many fields. In this chapter we shall meet it in four roles.

Compound interest

The first meaning of e, discovered by the Swiss mathematician Jacob Bernoulli (1654–1705) is about compound interest. Here is an example: customer A invests \$1000 in a savings plan that yields a 10 percent profit per annum. How much money will A have after ten years? The first answer that comes to mind is \$2000, since the profit is 10 times 10 percent, which is 100 percent. In actuality, though, A will have more. A 10 percent profit each year means that each year the principal is multiplied by $1\frac{1}{10}$ (or expressed as a decimal — by 1.1). After one year, A's account will contain 1.1 times \$1000, that is, \$1100. At the end of the second year he will have 1.1 times \$1100, which is \$1210. At the end of the third year he will have 1.1 times \$1210, and so forth. After k years, A will have $\$1000 \times 1.1^k$. To answer our question, after 10 years A will have $\$1000 \times 1.1^{10}$, and since 1.1^{10} is approximately 2.59, he will have about \$2590.

Customer B is more aggressive than A. He demands (and receives) interest of $\frac{1}{20}$ every half a year. How much will he have after 10 years? Before answering, think: will B earn more or less than A? The answer is — more. If it were not for this being compound interest, two half-years of $\frac{1}{20}$ interest would equal one year of $\frac{1}{10}$ interest. Since, however, this interest is compound, the interest for the two half-years comes to more than $\frac{1}{10}$, so each year B will earn more than A. Using a calculation similar to that for A's profits, 20 half-years will end with B having $1000 \times (1 + \frac{1}{20})^{20}$, which is about 2650 dollars.

Customer C is even more assertive. He demands that the 10 years be divided into 50 parts (50 fifth-years), and that he receive $\frac{1}{50}$ interest every fifth-year. The same calculation reveals that after 10 years the initial \$1000 investment will be multiplied by $(1 + \frac{1}{50})^{50}$, which is about 2.69. If a customer demands that the 10-year period be divided into 100 parts, and that he receive interest of $\frac{1}{10}$ each tenth-year, his money will be multiplied by $(1 + \frac{1}{100})^{100}$, which is about 2.704, so that after 10 years he will have a sum of approximately \$2704 in his account.

The sequence we are looking at is $(1 + \frac{1}{n})^n$. As we saw from the examples, the terms of the sequence increase as n increases. Yet, the rate of increase decreases, and therefore the sequence does not tend to infinity, but rather to a finite number, which is only a bit larger than 2.7. The limit of the sequence is marked by e, which is approximately 2.718. This value is only approximate, because e is not a rational number. Bernoulli's definition is the most commonly used of all the definitions of e.

It is not hard to see that if the numbers $(1 + \frac{1}{n})^n$ converge to e, then the numbers $(1 - \frac{1}{n})^n$ converges to $\frac{1}{e}$. To see this, write $1 - \frac{1}{n} = 1/(1 + \frac{1}{n-1})$, so

$$\left(1 - \frac{1}{n}\right)^n = \frac{1}{(1 + \frac{1}{n-1})^n} = \frac{1}{(1 + \frac{1}{n-1})^{n-1}} \times \frac{1}{1 + \frac{1}{n-1}}$$

The last term tends to 1, and the one before it to $\frac{1}{e}$. There is a gruesome story related to this calculation, told by the British writer and Nobel laureate Graham Green. As a teenager he suffered from acute depression, and he played Russian Roulette 6 times. In each attempt he had 1/6 chance of being killed. What was the probability of his survival? Since at each attempt he had a 5/6 chance to stay alive, the answer is $(\frac{5}{6})^6$, which is 5/6 of 5/6. . . of 5/6 (6 times), namely $(1 - \frac{1}{6})^6$. As we saw, this number is very close to $\frac{1}{e}$, namely a bit more than a third. Not much. Green's fans should be relieved, but of course they wouldn't know if he did succeed in killing himself.

Secret friend

Another instance of the appearance of e in real life is related to the custom of "Secret Friend" week in schools. Each child in the class is assigned a classmate, his or her "secret friend," who anonymously brings him gifts for an entire week. The assignment of secret friends is done by lottery: slips with all the children's name are placed in a hat, and each child picks a slip with a name of his or her secret friend. Obviously, there is a problem if a child takes a slip with his own name.

What is the probability that no child will take a slip with his own name?

Surprisingly, the answer is almost independent of the number of children in the class: the probability is almost exactly $\frac{1}{e}$. "Almost exactly" means that the larger the number of children in the class, the closer this probability is to $\frac{1}{e}$. The rate of convergence of this sequence is the fastest of all three examples considered so far — the numbers approach $\frac{1}{e}$ very quickly. If there are 30 children in the class, the probability differs from $\frac{1}{e}$ only after the 30th digit!

Geometric mean

The average weight of 5 people is the sum of their weights divided by 5. If we replace the weight of each one in the group with the average, the sum will remain the same.

Example:

The average of 1, 2, 3, 4, and 5 is 3, because
$1 + 2 + 3 + 4 + 5 = 3 + 3 + 3 + 3 + 3.$

This is also called "arithmetic mean." There are also other kinds of averages, the best known of which is the geometric mean. The geometric mean of a set of numbers is the number that will leave the *product* unchanged if it is exchanged for each of the numbers in the set.

Example:

> The geometric mean of 3, 8, and 9 is 6, because if we replace each of the three numbers with 6, the product will be $6 \times 6 \times 6 = 216$, which is exactly the product $3 \times 8 \times 9$.

The arithmetic mean of 3, 8, and 9 is one-third of $3 + 8 + 9$, which is one-third of 20, namely $6\frac{2}{3}$, a bit more than the geometric mean.

This is no coincidence. It is always the case: the arithmetic mean is greater than or equal to the geometric mean. The two averages are the same only if all numbers are equal: for example, the arithmetic mean of 3, 3, 3, 3 is (of course) 3, and this is true also for the geometric mean. Look now at the arithmetic mean of the numbers between 1 and 100. As was seen in the chapter "To Discover or to Invent," it is the middle between 1 and 100, which is $50\frac{1}{2}$. So, the geometric mean should be less. What is it? The answer is about $\frac{100}{e}$. In general, the geometric mean of the numbers $1, 2, \ldots, n$ is approximately $\frac{n}{e}$. This fact was proved by the Scottish mathematician James Stirling (1692–1770).

What does the word "approximately" mean here? It means that the ratio between the geometric mean and $\frac{n}{e}$ approaches 1 as n tends to infinity. In our example, in which n is 100, the geometric mean is approximately 37.993, while $\frac{100}{e}$ is about 36.788. The ratio between these two numbers (approximately 1.03) is quite close to 1. The higher the value of n, the closer this ratio will be to 1.

A differential equation

A ladybug is standing in the Cartesian plane (see the "Poetic Image, Mathematical Image" chapter), at the point (0, 1), that is, on the y axis, at a height of 1. It begins to move to the right, along a line with a slope of 1, that is, at an angle of 45 degrees with the axes, which means that for each unit that it moves to the right it will go up one unit. As the ladybug advances, it changes the angle of its movement — it always moves at an incline exactly equal to its height above the x axis. For example, when it reaches the height of 2 above the x axis, the incline of its movement will be 2, meaning that it goes up twice as fast as it moves to the right.

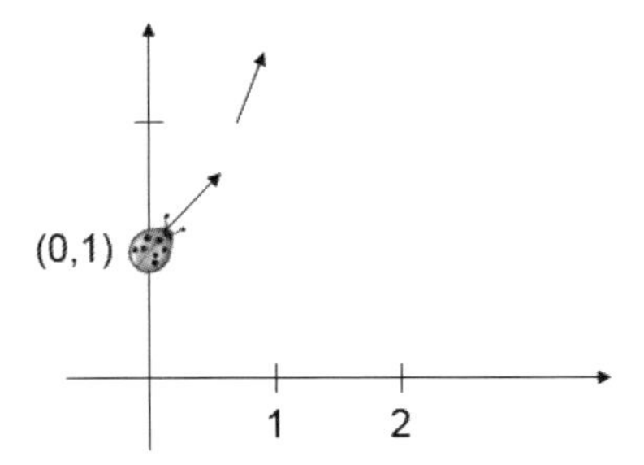

The ladybug is moving to the right and up. The ratio between its progress upwards and its movement to the right is increasing: it is equal to the ladybug's height above the x axis. It can be shown that when the ladybug has advanced x units to the right, its height is e^x.

Obviously, the ladybug will climb very quickly: the further it rises, the greater the rate of its ascent. The question is: what is the formula for the curve that it follows? That is, after it has moved x units to the right, how high will it be? The formula for this curve is $y = e^x$. In other words, after the ladybug has moved x units to the right, it will be at height e^x. For example, after a single time unit, it will be at height e; after two time units, it will be at height e^2. In terms taken from differential calculus, the function e^x possesses a unique trait, that it equals its derivative (the derivative of a function is its rate of change). This is the special property of the number e, from which all its other characteristics are derived. The function e^x does indeed rise very rapidly: for example — after 10 units of movement to the right, the beetle's height, e^{10}, will be approximately 6000 units.

What I have just described is called a "differential equation." Such an equation prescribes the behavior of a curve. It tells what the slope of the curve is, that is, the rate of increase of the value y as we advance along the curve, at every point on it. The differential equation above says: If this curve is described by the equation $y = f(x)$, then the function $f(x)$ is equal to its derivative. As a formula, this becomes $f'(x) = f(x)$. (The derivative of a function $f(x)$ is denoted by $f'(x)$.) The simplest differential equation is $f'(x) = 0$, which means: the derivative is 0, or, in other words, the rate of change of the function is 0 (more simply put, the function does not change). The solution of this equation is a constant function $f(x) = c$, where c is any fixed number.

Differential equations describe the geometry of curves, in terms of the rate of their change. They are one of the most useful mathematical tools.

Leonhard Euler (1707–1783), the greatest mathematician of the eighteenth century, and one of the most prolific mathematicians of all time. The blindness from which he suffered for the last seventeen years of his life did not curb his creativity.

Reality or Imagination

Don't trust them too far

Some numbers look real, and others less so. Some appear as if they are part of the real world, while others seem to be an arbitrary invention. The most natural are the numbers that are indeed called — "natural": 1, 2, 3, ... They are visible and tangible. Four apples are concrete reality. Fractions, too, never suffered from a lack of faith in their existence, since a half or a third of something too can be seen and felt. This is not to say that they reached their present form smoothly. The Egyptians knew only fractions with a numerator of 1, such as $\frac{1}{2}$, $\frac{1}{3}$ or $\frac{1}{4}$. The Romans had no special symbols for fractions, only verbal descriptions. The fraction sign that we know was invented by the Indians, and was brought to Europe only in the twelfth century by the Arabs. But despite their difficult birth, the existence of fractions was never doubted.

In contrast, there are some numbers that were, and are still suspected of being fictitious. It seems that they are no more real than the unicorn, and that the mathematicians who study them are playing make-believe. This was the fate of the number 0, that, at least on the face of it, has no object, that is, it doesn't count anything. In Europe it gained respectability only in the twelfth century.

Something similar happened to the negative numbers. When present day elementary school children recite a descending numerical sequence, say, 9, 7, 5, 3, 1, there are usually two or three children in the class who would continue -1, -3, -5. This just goes to show how natural the concept of negative numbers is today. Children pick it up as if from the air. It is hard to believe that this idea was still ferociously attacked in the nineteenth century. The famous French mathematician Lazare Carnot (1753–1823) argued that "in order to think about a negative quantity, something must be subtracted from 0, that is, from something that does not exist — which is impossible." The author of a mathematic handbook, Busset, argued that the root of the problem in mathematical education in France was teaching negative numbers. He wrote that "thinking about quantities smaller than zero is the height of insanity." In 1831 Augustus De Morgan, an important English mathematician, wrote that the appearance of negative numbers in real-life problems is just an indication of incorrect formulation. If we ask how much a store owner earned, and the answer comes out to be -10, it means only that we should have asked: "How much did he lose?", and then the answer would be the positive number 10. If we were to ask, "In how many years will Robert be twice as old as Sherman?", and we get -3, we should have asked: "How many years ago was Robert twice as old as Sherman?", and then the answer would have been a positive number: "3 years ago." Today, the idea that "a profit of -10 dollars" means a loss of 10 dollars, or that "in -3 years" is simply 3 years ago, is almost self-evident.

Other numbers were thought to be even more imaginary. They were even called this: "imaginary numbers." Today, these numbers have similar standing to the real numbers, but are still cloaked in an aura of mystery. Indeed, there is something magical about them.

Imaginary and complex numbers

The need for new types of numbers usually arises in the solution of equations. Negative numbers are necessary to solve equations such as $5 + x = 2$ (whose solution is $x = -3$); rational numbers were introduced to solve equations such as $3x = 2$ (whose solution is $x = \frac{2}{3}$). Irrational numbers first appeared in order to solve equations like $x^2 = 2$, that is, in order to be able to refer to $\sqrt{2}$. The next class of numbers, the imaginary numbers, also can be understood in this manner: they are needed to solve equations of the form $x^2 = -9$. The square of a real number cannot be negative, so there is no real number satisfying this equation. For, $3^2 = 3 \times 3 = 9$, and also $(-3)^2 = (-3) \times (-3) = 9$. The discoverer of imaginary numbers (for, it was a discovery — a notion was waiting there to be discovered) was the Italian Rafael Bombelli (1526–1572). He spoke of a number whose square is (-1), that is, a number that is $\sqrt{-1}$, but he did not assign it a name. Nor did he call it an "imaginary number," a name derisively given it by Descartes. The accepted letter for $\sqrt{-1}$, which is i (the first letter of "imaginary" or the Latin "imaginarius"), was given it by Euler in 1777. Once we have this number it is possible to solve equations such as $x^2 = -9$: the two solutions are $3i$ and $-3i$.

In order to solve equations that are a bit more involved, such as $x^2 + x + 9 = 0$, we need combinations of i and real numbers like $5 + 3i$ or $\sqrt{2} + \frac{2}{3}i$, and in general, $a + bi$, with a and b being real numbers. Such a number is called "complex," because it is composed of two parts, one real and the other imaginary.

The fundamental theorem of algebra

New types of numbers were defined one after the other, and in each phase an additional number type was needed for the solution of equations. Is there an end to this process, or will it continue indefinitely? Fortunately, the complex numbers are, indeed, the Promised Land. There is no need for further number types to solve new equations. This fact, that is known as "the Fundamental Theorem of Algebra," was proved by the 22 years old Gauss in 1799. The theorem states that the complex numbers are sufficient to solve any equation of the type known as "polynomial," such as $x^5 + 4ix^4 - 3x^3 + 10x + 1 + i = 0$. Gauss showed that every such equation has a complex solution, and that if the equation is of the nth order, namely the highest exponent of the unknown is n, then it has n solutions (the equation in the example is fifth-order, since 5 is the highest power of x in it).

Actually, this is true for more than polynomial equations. Almost every equation has a solution. This is Pickard's Theorem, which states that even if one were to concoct some wild equation that uses regular operations (such as multiplication, division, raising to a power, trigonometric functions), chances are that it has a complex-number solution. "Chances are" means that the equation has a solution for any value in its right-hand side, apart possibly from one single value. For example, there is a solution for the equation $2^x = 3 + i$ as there will be for any other number placed in the right side in place of $3 + i$, except for a single number: 0. There is no solution for the equation $2^x = 0$. "Almost every equation has a solution" is quite surprising. As far as the solution of equations is concerned, complex numbers are truly the last word. It is not that mathematicians stopped inventing new classes of numbers. But the new classes are not there for solving equations.

How complex numbers were born

Already the ancient Greeks spoke of the need to invent numbers with negative squares, but they did not devise any special letter for such numbers, nor did they give them the official status of "numbers." The story of the birth of the complex numbers, in the sixteenth century, is intriguing. The mathematicians of the sixteenth century simply brushed off quadratic equations such as $x^2 = -9$ as unsolvable, which averted the need to invent new numbers. But at that time mathematicians already were capable of solving third- and forth-order equations, and an embarrassing phenomenon emerged: at times an equation had real-number solutions, but complex numbers were needed in an interim stage of the solution. Since, in this case, they were not willing to forgo the solutions, the mathematicians of the time were forced to acknowledge the existence of numbers with a negative square.

As often happens, the seemingly purely theoretical entity proved extremely useful. Complex numbers have applications in wave theory, electrical engineering, quantum theory, and many other fields. Why is this so? Mainly because of a magic formula, discovered by Euler.

The world's most beautiful formula?

Mathematical beauty contests can be held by category: theorems, proofs, formulas. In the formulas division, there are no contestants to Euler's formula:

$$e^{i\pi} = -1$$

Note that in this short formula four of mathematics most important numbers appear: 1, π, e, i, as well as the minus sign. Euler's contemporaries were so enchanted by this formula that they expected it to reveal the secrets of the universe. This didn't happen, but the formula is unquestionably important, and is totally unexpected. Its real importance lies in it being a special case of a more general formula, also by Euler. This formula states that for every real number x (and, actually, for complex numbers, as well), the following holds true:

$$e^{ix} = \cos(x) + i\sin(x)$$

On the right side, an angle x is measured in "radians"; a radian is approximately 57°. The precise definition of a radian is: 2π radians (that is, approximately 6.28 radians) are 360°, which means that π radians are 180°. The formula $e^{i\pi} = -1$ is obtained by substituting π for x, since $\cos(\pi) = -1$

and $\sin(\pi) = 0$. The trigonometric functions $\cos(x)$ and $\sin(x)$ are the basic tools used to describe waves. This explains why the formula is so useful in physics and in electrical engineering, where waves play a central role. Euler's formula states that when applied to the complex numbers, the trigonometric functions and raising to a power are almost the same.

Unexpected Combinations

Eureka

Hiero, the king of Syracuse, assigned Archimedes (the greatest mathematician of the ancient world, 287–212 BCE) the task of checking the purity of the gold in the crown that had been prepared for him. Archimedes was soaking in his bath, when he suddenly made the connection between Hiero's problem and his sensation of weightlessness in the water. The words "Eureka! Eureka!" (I discovered! I discovered!) that the legend has him shouting when he ran into the street nude, have been used ever since for discoveries, whether earth-shattering or not. This discovery gave birth to the famous Archimedes principle, which states that the loss of weight of a body in water is the same as the weight of the displaced water, that is, the same as the weight of the part of the body immersed in a liquid, if this part were composed of the same liquid.

Arthur Koestler, the Jewish-Hungarian-English author of *The Act of Creation* argued that this is the only way creative ideas are born. Every creation comes into being by combining ideas. In other words, creativity is not *ex nihilo*, out of the blue, but connects existing ideas. As Koestler put this, it is the result of the meeting of two planes of thought. He called this process "bissociation." A classic example is the discovery of vaccine by Louis Pasteur. In one of his experiments, Pasteur infected chickens with cholera germs. One day, returning from his summer vacation and not wishing to throw away the culture of germs that remained from before the summer, he infected a batch of chickens with it. The chickens became only mildly ill, and quickly recovered. This wasn't surprising, since the germs were old and weak. The surprise was that when Pasteur's assistants tried to infect these chickens with regular cholera germs, they did not develop the disease. When Pasteur heard of this, he stared for a moment, and then said: "Don't you see? They were vaccinated!" This was not only one of the most useful flashes of insight in human history, but also one of the brightest, since the term "vaccinated" did not mean then what it means today. At the time the word "vaccine" (from "vacca," the Latin word for "cow") referred to a process discovered by the English physician William Jenner about a century earlier. Jenner heard from villagers that humans who had been infected by cowpox were immune to human smallpox, and recommended that humans be intentionally infected with cowpox. He knew nothing about the mechanism behind this process, and didn't dream about such creatures as germs. Pasteur understood this mechanism in a flash, linking Jenner's procedure with what had happened to his chickens.

Many scientists attest of themselves that they made their discoveries in this same way. Michael Atiyah, a great twentieth-century mathematician, tells that he would talk with people about their work, hear ideas from all different directions, and often things would connect with the problems that occupied him at that particular time. Richard Feymann explained how easy it is to become a genius. "It is easy," he testified. "I just roll in my brain the problem with which I am concerned at the time, until something comes and connects to it."

In a way, this is how the most famous conjecture in mathematics, named after Fermat, was solved. The conjecture (now theorem) stated that for any number n larger than 2, there are no integers x, y, z, all different from 0, for which $x^n + y^n = z^n$. This was also known as "Fermat's Last Theorem," since Fermat claimed that he had proved it. He used to write down his discoveries in the margins of his copy of Euclid's geometry book. He wrote this conjecture in the margin, and next to it: "I have a wonderful proof, but there isn't enough room to write it." The conjecture remained open for some 350 years, and tortured generations of mathematicians with the mocking simplicity of its formulation. A decisive step in the direction of its solution was the discovery of an unexpected connection with a conjecture that was formulated in the 1950s, and that seemed completely unrelated, the Taniyama-Shimura Conjecture. When the latter was solved in 1995 by the Englishman Andrew Wiles (with a proof that was too long for the margins of any book), Fermat's Conjecture was proved together with it.

Koestler's claim that every discovery is the result of joining together previous ideas is probably an exaggeration. A completely new idea, unrelated to known conceptions, has to appear every once in a while. But we cannot deny that this is one of the richest sources of beauty.

Unexpected combinations in poetry

The melody you abandoned for naught yet returns
And the eye of the way opens wide.

The appearance of unexpected combinations is the most expected occurrence in poetry. Take, for example, the above two lines, that open Nathan Alterman's book of poetry *Stars Outside*. The wording "eye of the way" perplexes the reader who must then search within himself: will going along the way open worlds for the walker, like the opening of one's eyes? Is the way to be found more in his eyes than outside? Here is another Alterman combination, the beginning of "Endless Encounter" (the second poem in *Stars Outside*):

> *For you stormed upon me, forever I will play thee*
> *For nought a wall will stop you up, for nought I will erect barriers!*

The storm joins the playing of music and the frothy river that nothing can withstand, and we realize that even a beloved one can be played. The poem then moves on to combinations such as "warring street, dripping raspberry syrup," "cities of trade, painful and deaf."

Unexpected combinations means bringing together distant patterns, and finding likeness between them. We saw that, in the eyes of many poets, this is the main characteristic of poetry. The poetical combination can be odd, but it is never arbitrary. It uncovers a true similarity between two patterns. Indeed, there is a likeness between opening one's eyes and the unfolding of the way as one walks. "Warring street," or "street dripping raspberry" are apt combinations, not because a street can fight or drip blood, but because the lover fights a lost battle, and he drips pain. In terms of the external world, there isn't much logic there, in terms of inner meanings there certainly is.

Harnessed together

You shall not plow with an ox and an ass together.

Deuteronomy 22:10

There is a term in Greek rhetoric known as "syllepsis," meaning joining far apart elements. Nathan Alterman was a master of such combinations.

From silence and glass panes
the nights of June are brittle

Nathan Alterman, "A Poem about Your Face," *Stars Outside*

The reader is forced to seek the meaning of these words within himself. Perhaps he will see in his mind's eye walking on a quiet street in front of windowpanes, or think how fragile is the quiet, and how precious. See, for example what happens in another poem by Alterman from *Stars Outside*, "The Marketplace in the Sunshine":

> *With the dust gathering, despotic, and bubbling tumult,*
> *With the fiery apples and in the storm of the oil,*
> *With the cry of the iron bending to the smith,*
> *With the thousand shields of the tin tubs,*
>
> *The marketplace stands,*
> *Foaming in the sun!*

Nathan Alterman, "The Marketplace in the Sunshine," *Stars Outside*

The richness of the marketplace demands a wealth of imagery. The poem speaks of the noise and tumult in terms of light, and then ascribes to it the human traits of a king. The fire is linked to apples (possibly referring to the shape of the flame); the iron "bends" (like before a king); the tubs are the shields of knights; the light joins a river's foam — there are many more than just two planes of thought here.

Nathan Alterman (1910–1970), born in Warsaw, Poland; immigrated to Palestine in 1925.

What is Mathematics?

The hard life of the definer

They say there is love in the world.
What is love?

Hayyim Nahman Bialik, "Take Me under Your Wing"

I shall not [...] attempt further to define [pornography].
[...] But I know it when I see it.

Potter Stewart, United States Supreme Court justice

We may not know quite what we mean by a beautiful poem, but that does not prevent us from recognizing one when we read it.

Godfrey Harold Hardy, *A Mathematician's Apology*

Ask a mathematician to define his profession, and chances are that he will stutter. A physicist can say what it is that he studies, but a mathematician, even after long years of research, will find it hard to define his occupation. One of the common definitions of mathematics relates to its subject matter: "the science of the number and shape." In other words, numbers and geometry. There is much truth to this. Almost every modern mathematical field developed from one of these two fields, and almost every mathematical topic has a geometric or numerical aspect. And strangely, the more mathematics advances, the harder it becomes to separate the two: geometry contributes to numbers, and numbers appear in geometry. But the reservation "almost" is unavoidable: numbers or geometry appear in only almost every mathematical field, not in all. For example, mathematical logic (which will be the subject of a later chapter) touches upon neither numbers nor geometry. Or take the ants puzzle in the first part: the concepts it used, — "collision," "change of direction" — were not numerical, and were hardly shape-related.

Abstraction

If so, then what is special about mathematics? To answer, let me tell a story from elementary school. When I go there I sometimes ask first-graders: how many are 2 pencils and 3 pencils? The children learned that addition means joining, so they join 2 pencils to 3 pencils, and find that they have 5 pencils. Now I ask: how many are 2 erasers and 3 erasers? They immediately answer, "5 erasers."

> "How do you know?"
> "Because we saw that with the pencils."
> "So what?" I argue. "Maybe it is different for erasers?"

The children laugh. But my question is serious. Behind it lies the main strength of mathematics: generality. Mathematics strips a situation of its secondary details, and leaves the gist. In this case, the gist is that three objects and two objects are five, regardless of their nature or arrangement in space. Of course, we make abstractions all the time, and in every field of thought. What is special about mathematics is that it takes abstraction to an extreme, applying it to the most basic thought processes.

The classic example is the concept of number. Numbers were born by the abstraction of the most fundamental thought process: the division of the world into objects — separating the world into units and assigning them names: "apple," "family," "state." Counting means repeating the same unit many times "2 apples," "3 apples," "4 apples". . . .

Frege

The industrial revolution in nineteenth-century Europe had far-reaching consequences, not only for man's quality of life, but also for his self-perception. Once it became clear that machines could replace human muscles and skills, the path was short to the idea that man himself closely resembles a machine. It is no surprise that Darwin's theory of evolution, which changed man's view of his place in the world, came into being at that time, and in England, the cradle of the industrial revolution. All this quickly led to the idea that a machine could replace humans in thought, as well. In the middle of the nineteenth century Charles Babbage attempted to build an innovative calculating machine. Indeed, the mathematician-philosopher

Blaise Pascal preceded him in building a calculating machine, but Babbage's machine had a wonderful feature: it could be programmed. Babbage did not complete his project (a machine constructed by his blueprint was made only a century and a half later), but the idea made waves in England.

About twenty years later, in the 1870s, a mathematician-philosopher named Gottlob Frege (1848–1925), from the University of Jena in Germany, conceived of an even more ambitious idea. He proposed that not only could a machine conduct mathematical operations, but also human thought itself behaves like a machine. And if thought is mechanical, then it can be studied mathematically, just like the movements of celestial bodies or the flow of liquid in a pipe. Of all human thought, it is mathematics that obeys the clearest and best-defined laws, and hence should be studied first. This was a staggering idea: investigating mathematical thought in a mathematical way. In other words, devising a mathematics of mathematics (or "metamathematics," as this is often called).

In the computer age, such an idea seems natural. Today we know that machines can think, and that their thinking can be the subject of mathematical study. At the end of the nineteenth century, when even Darwin's theory was fresh and controversial, this was a daring insight. It required man to surrender his last claim to uniqueness: abstract thought. In hindsight, this was a turning point no less significant than Babbage's calculating machine.

Mathematical thought has many facets: constructing concepts that correspond to phenomena in the world, forming conjectures, and building theories. But there is a single part that mathematicians value most: proof. Frege limited himself to this aspect of mathematical thought, mainly because it is amenable to formalization. He argued that a proof is a mechanical process, a game with symbols on paper, that obeys clear, and even quite simple rules. It is a "game" in the sense that it has rigid rules, that are not essentially different from, say, those of chess. There are permitted moves, and there are forbidden ones. Frege defined a proof as a series of sentences written (let us say) on paper, each of which ensues from its predecessors, by one of a limited number of deductive rules. It is easy for us, living in the twenty-first century, to accept this idea, because this is the way that a computer works. A computer thinks by taking a series of symbols encoded in electrical signals, and acting on them in accordance with the rules programmed into it. It receives as input a series of symbols, and produces as output another series of symbols. Today, all this is self-understood; in Frege's time, understanding that a formal operation on symbols on paper could be considered as "thought" was a breakthrough.

Frege's ideas appeared first in a paper, and then in a book published in 1879 that was intended to be the first of three volumes on the foundations of arithmetic. The book was almost totally ignored. The only reaction was a devastating review by Georg Cantor (whom we already met in the chapter "The Real Numbers"). As we will see, Cantor himself, in his turn, would receive similar treatment by other mathematicians. The embittered Frege published a second volume, in which he attacked the mathematicians of his time. An intended third volume never appeared.

It might have taken humankind even longer to digest Frege's discoveries, but for the fact that Bertrand Russell had as a child a German nanny, and therefore knew German. Russell traveled to Germany, read Frege's articles, and understood their importance. Once he returned to England he enlisted the help of his teacher at Cambridge, Alfred North Whitehead, to a mega task: writing parts of the mathematics of the time in the formal language devised by Frege. The result of their work, *Principia Mathematica*, was a heavy, almost unreadable, book, but changed the course of the subject called "mathematical logic."

For Frege, mathematics means playing with symbols. The mathematician chooses a system of axioms, such as the axioms of number theory (for example: "For every number n, $n + 0 = n$"). Then, he examines which theorems can be proved from the axioms, in accordance with rigid and predetermined rules of proof. As I already mentioned, this is a narrow perspective of mathematics. It ignores, for example, the step of selecting the axioms: why was a particular system chosen, and not another? Systems of axioms do not just spring out of the blue. They are meant to describe reality. The quality of the system of axioms will determine whether they will lead to profound conclusions, or will be barren. As experience witnesses, if the axioms come from reality, they will prove to be fertile.

Frege tells us that the correctness of a proof can be checked mechanically. And how about the much more important task, of finding proofs, namely the act of proving itself? As we shall soon see, there is no mechanical way of doing this. There are no recipes for proving theorems. And as of today, computers lack the intuition that leads mathematicians to their proofs. I personally believe that one day this aspect of mathematical thought, as well, will be understood, and computers will be capable of proving theorems, formulating conjectures, and, in short — replacing mathematicians in all aspects of their work. And when this happens, it will be thanks to Frege and his work.

Gottlob Frege (Germany, 1848–1925), mathematician and philosopher.

Deep Tautologies

It is like a déjà vu all over again.

Yogi Berra, baseball player

The vast majority of our imports come from outside the country.

George W. Bush

I always thought that record would stand until it was broken.

Yogi Berra

All mathematics is tautology

Ludwig Wittgenstein, Austrian philosopher, 1889–1951

What is the difference between mathematics and philosophy? In mathematics, someone important is someone who said something important. In philosophy, something important is something that somebody important said.

Anonymous mathematician

In 1820 Gauss suggested that in order to communicate with intelligent inhabitants of the moon, if there are such, forests should be cleared in Siberia to create a picture of the Pythagorean theorem (the same drawing that appears in the chapter "Mathematical Harmonies"). This theorem is valid everywhere in the universe, it was always there, and Pythagoras only had to discover it. In other words, its information is hidden in its assumptions. Seemingly, it does not really contain any new information, namely, it is a tautology. This is what Wittgenstein claimed in the above quotation.

"Tautologia" in Greek means "an identical word." It is an empty statement that contains no new information. "Water is wet," for example. A "tautological argument" in logic means an argument that is always true, such as: "If today is Tuesday, then today is Tuesday." By what we saw in the preceding chapter, a mathematical proof is a chain of tautologies. Each line in it ensues, aided by tautologies, from the preceding lines. This, however, is still distant from the conclusion reached by Wittgenstein in the quotation above, that mathematics is a collection of tautologies, because the challenge lies in connecting these tautologies in the right order. The correct theorem must be guessed, and the right lines to be added must be found. The Pythagorean theorem is indeed a combination of tautologies, but this doesn't mean that it is void of information, because it is an intelligent combination. "A proof is a nontrivial combination of trivial statements," as one of my teachers once said. As we all know, everything is easy after its discovery. Once the proof is given in full, it is child's play to check its correctness. The work lies in discovering it — that is, knowing in which direction to go. To say that "mathematics is tautological" is like saying that the Eiffel Tower is nothing more than a collection of metal rods and screws.

Tautologies with a message

Something similar happens in poetry: there, too, apparent tautologies often convey hard truths. Poetical tautologies are written with a wink: the emptiness is misleading. The next poem, by Nathan Zach, "I Hear Something Fall," is a chain of tautologies.

I hear something fall, said the wind.
Nothing, this is just the wind, the mother reassured.

You are guilty and you, too, are guilty, the judge
ruled for the accused. Man is only a man,
the doctor explained to the stunned relatives.

But why, why, the youth asked himself,
not believing his eyes.

Those who don't live in the valley live on the hill
the geography teacher said
with no special zeal.

But only the wind that let the apple fall
remembered what the mother hid from her son:

That no consolation will ever, ever come.

Nathan Zach, "I Hear Something Fall," from *Other Poems*

Amusing? Perhaps. But it is also hard to miss the despair. The heroes of the poem face the empty statements dumbfounded, and the message is that man's very existence is filled with despair; that nothing specific needs to be known in order to realize the finality of life.

Haiku poems

Haiku is a strict Japanese poetic style, with exactly 17 syllables (in the original). Haiku poems frequently depict a nature scene that relates to a specific season. The Japanese poetry scholar R. H. Blyth said that haiku poems are always tautological. They contain no new information. Haiku poems are more about what does not happen than about what does:

Darkening sea
voices of the wild ducks
faint whitely away.

Matsu Basho, 1644–1694

Externally, the ducks have gone. Internally, something has happened: their voices left their mark upon the listener. The minimalism of the outside action leaves a place for inner action.

Arriving alone
to visit someone alone
in the autumn dusk.

Yosa Buson, 1715–1783

The wording "Arriving alone to visit someone alone" is almost a tautology. But who visits whom is not important — what is significant is what happens within oneself, and the dusky autumn sensation that arises upon reading the poem.

The old pond;
A frog jumps in —
The sound of the water

Matsu Basho, trans. R. H. Blyth

The frog, like the ducks, vanished. And yet another tautology: the sound of the water, Blyth writes, is subsumed within the old pond — there is nothing new here.

Dawn rises
the storm is buried
in shrouds of snow.

Edo Watsujin, 1758–1836

Yet the day after the storm
the redness of the pepper.

Matsu Basho

These poems tell us: "the tempests beyond are unimportant," and what happens within is not dependent on the external events, as it would seem.

A camelia blossom fell.
A rooster crowed.
Another blossom fell.

Sakurai Baishitsu, 1768–1852

Symmetry

Tyger! Tyger burning bright
In the forests of the night
What immortal hand or eye
Could frame thy fearful symmetry.

William Blake, "Songs of Innocence and Experience"

Economy by symmetry

In a schoolbook that I had as a child there was a story about a king who assigned to two artists the task of painting the royal chamber, one side of the chamber to each. One artist labored for months, while the other just sat idly. On the last day, after the first artist had completed his work, the other artist placed mirrors on his side. The king came, and was greatly impressed by the paintings on the first side. When he came to the other side, the second artist showed him that his side had just as fine paintings. The king paid the first artist his wages, and said to the other: "Here, you see the bills in the mirror? You can take them on your side."

Mathematicians are very fond of the second artist's stratagem. Mathematics never spares effort in order to spare effort, and it often, and very successfully, uses symmetry for this purpose. There is no need to work twice when a single time will do. Here is a famous example:

> Two people are situated on the same side of stream L, one at point A and the other at point B. This is a mathematical stream, namely, it is a straight line. The person at B is dying of thirst, and the person at A wants to bring him water from the stream, as quickly as possible. He has to go to L, and from there to B. What route will he choose, so that he will have the shortest distance to go? That is, what is the shortest (broken) line that touches L and connects A with B?

In order to answer the question, we will reflect B with respect to L, which gives us point B′.

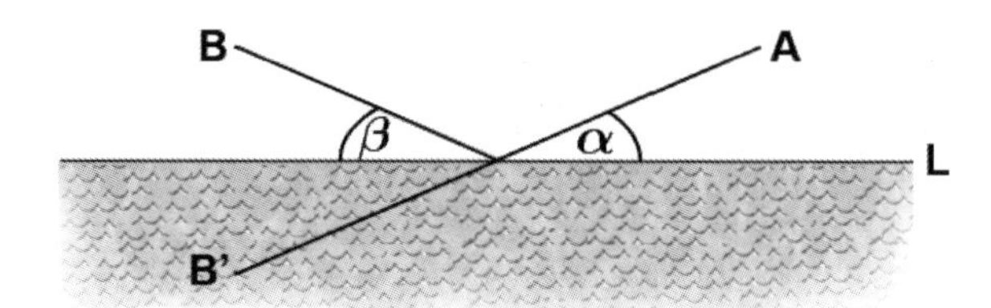

The shortest broken line from A to B is obtained by reflecting the stream part of the line from A to B′, the reflection of B.

Every line that connects A with B and meets L corresponds to a line of the same length that connects A with B′. The shortest line connecting A with B′ is of course the straight line. And so, in order to find the desired line between A and B, connect A to B′ with a straight line, and reflect with respect to L the part of the line on the side of B′. Note that on the line we obtain, the so called "angle of incidence" α equals the "angle of reflection" β.

This is useful in physics. The principle of minimization of energy states that a light beam passes between two points (even if reflected from a mirror) along the shortest path — nature economizes.

The way to build an arch is another example of economy by reflection. As is common knowledge, the Romans were no great innovators, and they borrowed elements of their culture from the peoples that they conquered, especially from the Greeks. They took the use of arches in building from the Etruscans, the inhabitants of Italy before them. This was an amazing invention: self-supporting bricks, that do not need cement to bind them together. In the ideal arch the pressure on each of the bricks is as small as possible, which means that the pressure is distributed evenly among the bricks. In a circular arch, which is probably the most natural and most common type, this condition is not fulfilled, because the stones on the sides of the arch are subjected to greater pressure than those in the middle. So then, what is the correct shape for an arch? This can be determined by calculations that are not especially complicated (the tool needed is differential equations). But there is a simpler method, that avoids calculations: leaving this to nature. Instead of building an arch, hang a string. Take a string of the length of the desired arch, and hang it from two points as far apart as the distance between the arch ends. The string will naturally assume a shape in which the tension is equally distributed. Now all that remains is to turn the string over, so that its arch points up. The Spanish architect Antonio Gaudi used this trick to plan his buildings.

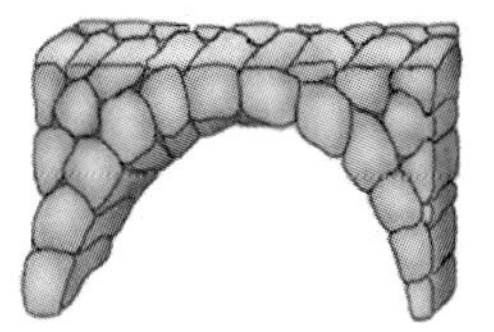

The round table game

Here is a game for two players. The players sit at a round table, and each has an unlimited quantity of same size coins. Each in turn places a single coin in an empty spot. The winner is the player who places the last coin, that is, after his move there is no room left for the other player to place a coin. Which player can assure a victory?

The answer is that the opening-move player has a strategy for winning. He places a coin exactly in the center of the table. After this, he adopts a mirror (or monkey) strategy: for every coin that the other player places anywhere, he places a coin symmetrically, on the opposite side. This strategy ensures that after each move the coin situation will be symmetrical, that is, every coin has a matching coin opposite it. This guarantees the first player a win, because he always can make a return move: since the coin placement is symmetrical, if there is a spot open for the second player, there will also be an (exactly opposite) opening for the first.

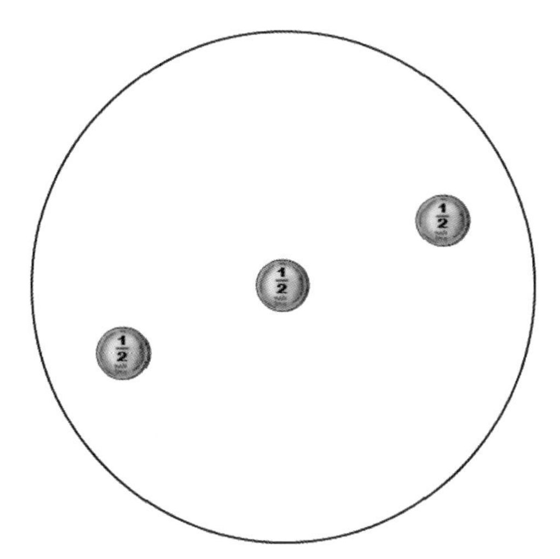

The first player places a coin in the center. Now, for every move by his opponent, he will place a coin exactly opposite.

The second player cannot use the mirror strategy, because of the unique opening move of the first player, that cannot be mirrored: there is no symmetrical response to placing a coin in the center. By the way, the table doesn't have to be circular. This same strategy works for a rectangular or elliptical table, because they, too, have a center.

Dido's problem

> A shepherd is given a rope, and is told that he can use it to fence off an area as he wishes. Which shape should he choose, in order to fence off as large an area as possible?

This is known as the "isoperimetric problem," since it seeks a shape of maximal area, given its perimeter ("isoperimetric" means "equal perimeter"). Another common name is "Dido's Problem," after the first queen of Carthage. Dido's brother, the despotic ruler of the city-state of Tyre, murdered her rich husband and expelled her as well. With companions, Dido landed on the shore of present-day Tunisia. For a sum of money, the tightfisted locals promised to give her a patch of land that she could fence off with the hide of one ox. Dido cut the hide into strips, which she joined together into a single chord. She could use this chord to fence off an area, one side of which was the sea shore. The question is: what shape should she have chosen for this area? Dido made the right choice: a semicircle. This is half of the shepherd's problem with which we began this section. The solution to the shepherd's problem, as can easily be guessed, is the shape with the greatest symmetry, namely, a circle.

Before explaining why this is so, we should first look at a simpler problem. Let us assume that the shepherd doesn't want to build his pen in just any shape, but as a rectangle. Obviously, different types of rectangles can be chosen, such as a very high and very narrow rectangle, or one that is very wide and very low. Which one should the shepherd select? If we take "very narrow" to an extreme, that is, a rectangle of width 0, its area will be 0; and also if we take "very low" to an extreme, its area will be 0. We can imagine that the optimal solution is in the middle between these two extremes, that is, a square. This is indeed the case. The shepherd should mark off a square with his rope.

Here is a simple proof. Let L stand for the average length of the sides of the rectangle. That is, L is the perimeter of the rectangle (which is the length of the rope) divided by 4. So, the perimeter is $4L$. The sum of the lengths of two adjoining sides is half the rectangle's perimeter, which is $2L$. Assume that the length of one of the two adjoining sides is $L + X$. Since the sum of the lengths of the adjoining sides is $2L$, the length of the other side is $L - X$ (since $L + X + L - X = 2L$). The area of the rectangle is the product of the lengths of the adjoining sides, which in this instance is $(L + X) \times (L - X)$. When we open the parentheses, we obtain $(L + X) \times (L - X) = L^2 - X^2$. But X^2 cannot be negative. Accordingly, $L^2 - X^2$ does not exceed L^2, which is the area of the square with the same perimeter (each of whose sides is L long). The smaller X is (that is, the closer the sides' length to each other), the larger the area of the rectangle.

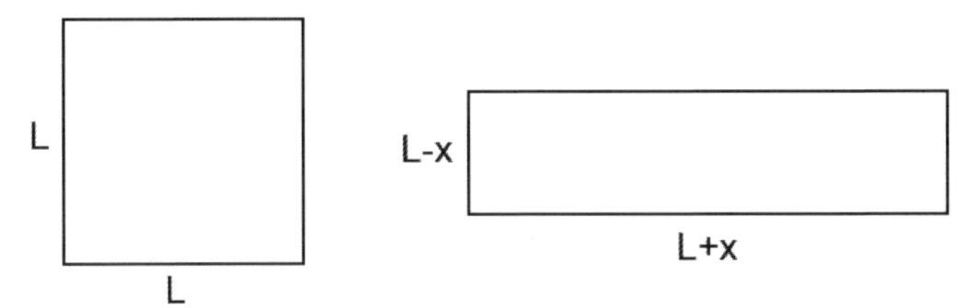

The square and the rectangle in the picture have the same perimeter ($4L$). The area of the square is greater.

Assume now that the shepherd still has to choose a rectangle, but he can use the bank of a stream as one of the boundaries of his pen (just as in Dido's Problem). As usual, we will assume that this is a mathematical stream, that is, a straight line. What shape should the shepherd choose for his rectangle? Here, too, formulas could be used, but it is simplest to rely on symmetry. The shepherd's area is mirrored across the stream bank line.

The combination of the original rectangle with the reflected one is the rectangular area that does not make use of the stream. Its entire perimeter is a rope (albeit one that is partially imaginary) twice the length of the shepherd's rope. Consequently, the new (doubled) rectangle's perimeter is fixed; and as we already know, its area is maximal if it is a square. Since the area of the shepherd's original rectangle is half of the overall rectangular area, the shepherd would do well to choose half a square.

Back to Dido and the isoperimetric problem

In the original (isoperimetric) problem, the shepherd is not limited in his choice of shape for his pen, and he can choose any shape he wishes. As we already mentioned, then he should choose a circular area. This argument is called the "isoperimetric inequality":

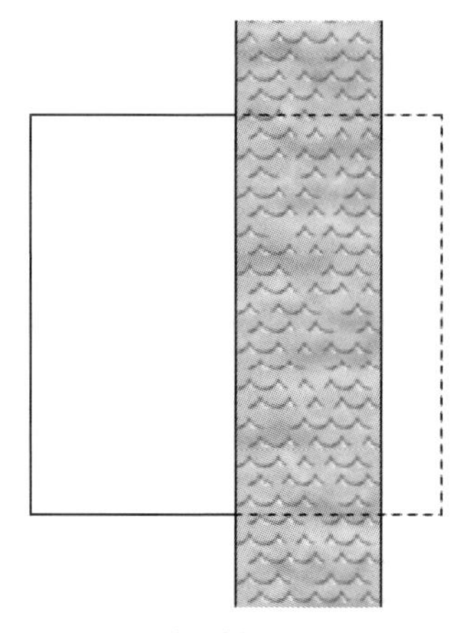

The shepherd wants to use his rope to build a rectangle to the left of the stream, with the stream itself being one side. The reflection of this rectangle (the dotted line) yields a rectangle with a perimeter twice as long as the rope, and an area twice that of the original rectangle. Of all the rectangles with a perimeter twice the length of the rope, a square has the largest area. Therefore, the shepherd should choose a rectangle that is half a square.

Of all the shapes with a given perimeter, the circle has the greatest area.

This theorem also has a three-dimensional version, which, too, is intuitive, but harder to prove: of all the bodies with a given surface area, a sphere has the largest volume. This is one of the reasons why mammals' heads are quite round. Why doesn't man utilize this, and build more spherical structures? Among other reasons, because spheres cannot be packed perfectly. It is much easier to pack rectangles and boxes. This is why rooms in houses, for example, have right angled corners.

But let us return to the two-dimensional case. It is not difficult to guess that the circle is the optimal shape, it is harder to prove this. While the theorem was already guessed by the ancient Greeks, an (almost) rigorous proof was given only in the middle of the nineteenth century, by the Swiss Jacob Steiner (1796–1863), a contemporary of Gauss. Steiner was an autodidact, and said that he hated formulas (he claimed that they hide the ideas) and loved geometry. He used the concept of symmetry, but in the direction opposite to the one used above in the rectangle case. In the case of the rectangle, we concluded from the whole regarding its half; now, following Steiner, we shall go from the part to the whole. Like the artist in the story, we will solve half the problem, and derive our solution for the whole from it. As in Dido's story, the shepherd will be told that he may mark off a pen with his rope (of fixed length), and that one side of the pen will be the bank of a river.

It is best for the shepherd (like Dido) to choose a semicircle. To show this, assume that the shepherd used his rope and the line L to mark off a pen of maximal area. A and B will be the points of contact between the rope and L (see the drawing).

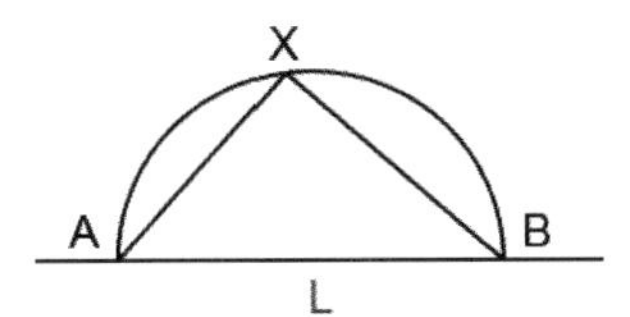

If the rope bounds a maximal area with the line L, then every point X on the rope "sees" the segment AB at a right angle.

We have to prove that the shape chosen by the shepherd has to be a semicircle with A and B as its ends. To show this, we will make use of a well-known geometric fact: when a semicircle of diameter AB is constructed, every point X on the semicircle "sees" the segment AB at a right angle. In other words, for any point X on the semicircle, AXB is a right angle. What is important for our purposes is that the opposite is true as well: any point that sees the segment AB at a right angle and is located above this segment is located on a semicircle the diameter of which is AB. Accordingly, all we have to prove is that every point on the rope sees the segment AB at a right angle, which will show that the rope is arranged in the shape of a semicircle. All this, of course, based on the assumption that the shepherd marked off a maximal area.

In order to understand the connection between the angle and the area, let us conduct a simple experiment, Extend both arms straight out (without bending them at the elbows), with some angle between them, and think of a triangle whose points are your head and your two hands. What angle should you form with your arms to form a triangle with the greatest possible area? If you spread your arms completely to the sides, that is, at an angle of 180°, the triangle will be completely flat, and its area will be 0. If you raise both arms straight up, and parallel (that is, at an angle of 0°; if your body didn't get in the way, they would be the same line), the area of the triangle, once again, will be 0. It is not hard to prove that the best case has your hands perpendicular, that is, at a right angle. The proof of this fact uses the formula for the area of a triangle as "the base times the height, divided by 2." Formally stated, given the sides XA and XB (the "arms") whose vertex is X (the "head"), in order to create the triangle of maximal area whose points are A, B, and X, the sides must meet at a right angle.

Let us assume that there is a point X on the curve for which the angle AXB is not a right angle (that is, not 90°). Now modify the curve by changing this angle to 90 degrees, without changing the two arcs over AX and over BX (see the next drawing). According to the argument, the area is larger — in contrast with the assumption that the curve bounds with L a maximal area.

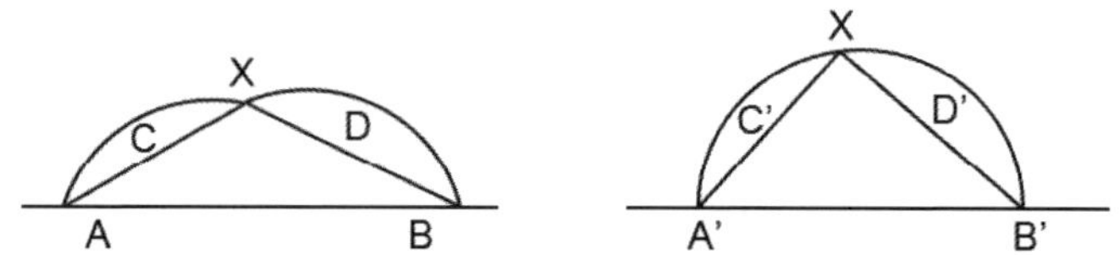

The right-hand picture is derived from the left-hand one by closing the span between the sides XA and XB, so as to make them perpendicular. The domes C and D are preserved. The length of the curve (the rope) is not thereby changed. The total area increased, because the area of the triangle increased, while the area of the two domes remained as it was.

Having worked hard to solve the riverbank case, we can use this to solve our original problem, of the shepherd who does not have a stream that can bound part of his area. In order to show that it is best for him to mark off a circular area with his rope, we will assume that he chose an area of some other shape. Take two points on the rope, A and B, that divide the rope (that forms a closed shape) into two parts of equal length (see the left-hand drawing):

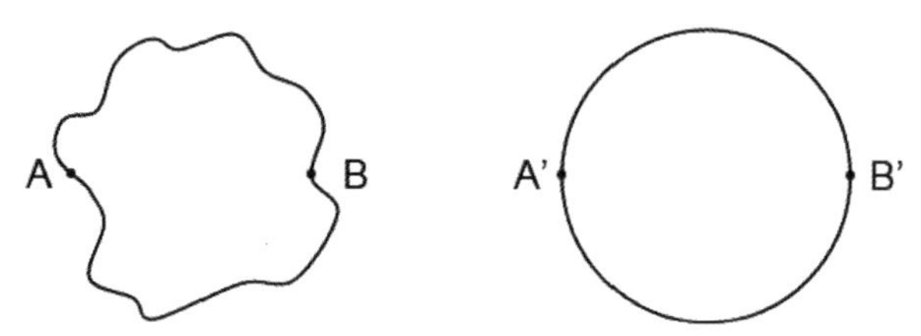

The two shapes have the same perimeter. The circle on the right was obtained by replacing the upper and lower parts (both of which are between A and B) in the left-hand shape by semicircles. According to what we already proved, this increases the area, and therefore the area of the circle is larger than that of the left-hand shape. This is the isoperimetric inequality.

Replacing both the bottom and top parts of the shape with semicircles will increase its area. To see this, draw a line between A and B, and note that by the riverbank case, each part of the shape cut by this line increases in area. But this is tantamount to replacing the entire shape by a circle — and this is exactly what we wanted to prove.

Symmetry in poetry

An old man, what has he in his life?
He wakes in the morning, but morning doesn't wake in him.

David Avidan, "A Sudden Evening," *Pressure Poems*

These are the opening lines of a poignant poem on old age by a 28-year-old. David Avidan, the *enfant terrible* of Hebrew poetry, was born in 1934. When he died in 1996, alone and penniless, it was hard not to recall this poem. I quoted it here because of the second line, that uses a poetic device called "chiasmus": the words "wake in the morning" are exchanged, to produce "a morning that (doesn't) wake in him." The term chiasmus has its source in the Greek letter χ, called "chi" (with the German guttural *kh*). In this example, the chiasmus proceeds from the outside in, from the morning outside to the morning (that is not) inside. The same happens in the continuation of the poem:

He shuffles to the kitchen, and there
the lukewarm water reminds him
that at his age, that at his age, that at his age,
an old man — what has he in his mornings?
He arises on a summer morning, and already fall
mixes with evening in the light bulb above.

Once again, the outside is reflected within: the lukewarm water reminds the old man of his lukewarm blood and his lukewarm life. In the last line there is mirror reflection, but of opposites: the summer without is reflected as autumn within.

In Nathan Alterman's powerful poem "The Foundling," inner reality is the mirror reflection of the external one. Externally, the mother abandons her baby; her inner truth is that he has abandoned her. The incongruity between the picture and its reflection in the mirror only grows throughout the poem. The symmetry of roles is joined by a temporal symmetry: the end, the mother's death, is a mirror image of its beginning, the son's birth; and her shrouds are a mirror image of his infant clothes. Nathan Zach, the other great "Nathan" of Hebrew poetry, who knew Alterman well, testified that the poem painfully reflects Alterman's relationship with his parents. I have selected three stanzas from this long poem:

At the foot of the fence my mother placed me,
face creased and still, on my back.
And I looked at her from below, as from a well, —
Until she fled as one flees from a battle.
And I looked at her from below, as from a well,
And a moon was raised over us like a candle.

[...]

She grew old in my prison and lean and small
And her face became creased as my face.
Then my little hands clothed her in white
Like a mother clothing her living child.
Then my little hands clothed her in white
And I carried her off without telling her where.

And at the foot of the fence I placed her
Watchful and still, on her back.
And she looked at me laughing, as from a well,
And we knew that we ended the battle.
And she looked at me laughing, as from a well.
And a moon was raised over us like a candle.

Nathan Alterman, "The Foundling," based on a translation of B. Harshav, *The Modern Hebrew Poem Itself*, eds. S. Burnshaw *et al.*

Impossibility

Ruler and compass constructions

One of the greatest contributions of the Greeks to geometry, along with the ideas of "theorem," "proof," and "axiom," was the concept of construction of geometric objects using only a ruler and compass. The ruler used for such constructions has no markers of length, and therefore cannot be used to measure distance, but only to draw straight lines between given points. A compass is used to draw circles, and to draw equal segments (a segment is a finite part of a line). While lengths cannot be measured using only ruler and compass, lengths of segments can be compared, and a segment equal in length to a given segment can be marked off on a line.

These do not seem to form an impressive arsenal of tools, but this is misleading. In actuality, much can be accomplished using these two simple devices. A parallel to a given line can be drawn, through a given point; a perpendicular to a line can be drawn from a given point on or outside it; an angle can be bisected; and a segment can be divided into any finite number of equal parts. It is possible to construct a perfect hexagon (with equal sides and equal angles), an octagon, and — as Gauss proved at the age of 19 — even a perfect 17-sided polygon. Gauss was so proud of his discovery, that was almost certainly the first significant geometric construction since the time of the Greeks, that it led him to prefer mathematics to philology, his other academic interest. He asked that a 17-sided polygon be inscribed on his tombstone. The tombstone mason refused, claiming that it was impossible to tell the difference between this shape and a circle. This injustice was corrected fifty years later, on a monument erected in memory of the great man.

When we speak of geometric constructions, we mainly think of finding points, or of forming shapes. But there is another type of construction, that of lengths. We are given a certain segment, that serves as the "measuring rod," namely, it is arbitrarily defined as 1 unit long. We then want to build a segment of length 2 (that is, double the length of the given segment), or of

$\frac{1}{2}$ (half of the given segment), and so on. It is easy to multiply the length of the segment by a whole number (by simply adding more and more copies of the segment, one next to the other), and it is not difficult to divide a segment into a whole number of equal parts. So segments of any rational length can be constructed (in order, for example, to construct a segment of length $\frac{3}{5}$ multiply the segment by 3, and then divide the result into 5 equal parts). Euclid, in the fourth century BC, already knew how to geometrically extract a root. That is, given a segment of length a, he could also construct a segment of length $\sqrt{a}$. This is done, as in the drawing below, by constructing a right angle triangle with a hypotenuse of length a, and such that the projection of one of its sides on the hypotenuse is of length 1. In the drawing, the length of the projected side is labeled x, and the length of this side is $\sqrt{a}$, that is, $x = \sqrt{a}$. Why? The triangles ACB and ADC have equal corresponding angles. This implies that they are similar, meaning that the ratios between their sides are equal. So, $\frac{AD}{AC} = \frac{AC}{AB}$ or, in other words, $\frac{1}{x} = \frac{x}{a}$. By multiplying the sides of the equation by the denominators (x and a), we obtain the equality $x^2 = a$, meaning that $x = \sqrt{a}$.

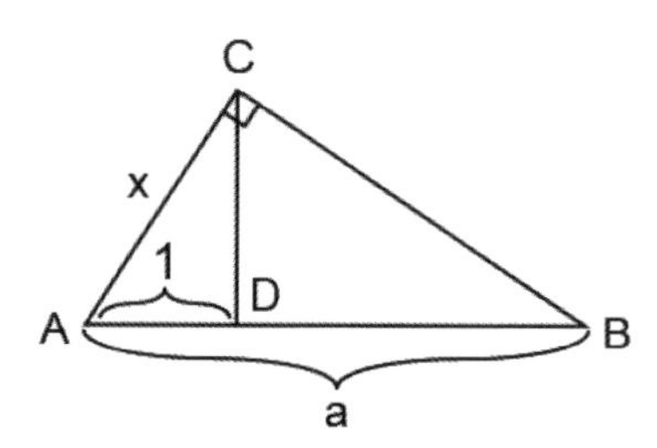

This construction can be used when a is larger than 1. The case of a being smaller than 1 requires another drawing, in which the roles of 1 and a in the triangle are interchanged, namely, the length AB is taken as 1, and the length AD is a.

Three impregnable fortresses

The power of the ruler and compass is indeed surprising. But three construction problems remained unsolved, and frustrated the efforts of both professional and amateur mathematicians for more than two thousand years. They withstood vigorous assaults first by the Greeks, and then by the mathematicians of Renaissance Europe. As is usual for famous problems, many erroneous proofs were proposed for each.

(1) The most famous of the three was **squaring the circle**: given a circle of a certain radius (say, one unit), construct, with the aid of a ruler and compass, a square with the same area as that of the circle. In a second formulation: find a segment whose length equals the circumference of the circle. The first to tackle the problem was Anaxagoras (499–428 BCE). Aristophanes derided the circle squarers in his play *The Birds*, and ever since "squaring the circle" has been synonymous with attempting to achieve the impossible.
(2) **Doubling the volume of a cube**: in the chapter "Mathematical Harmonies" we saw how, for a given square, it is possible to construct a square with double the area. This is plainly a square, the side of which is the same length as the diagonal of the original square. Can something similar be done for volume? Namely, for a given cube, can the side of a cube of double the volume be constructed with the aid of a ruler and compass?
(3) **Dividing an angle into three equal parts**: bisecting an angle using ruler and compass is simple. Can an angle also be divided into three equal parts?

Thousands of years of failure should have alerted mathematicians to the existence of some inherent difficulty. But the notion of impossibility of mathematical tasks did not emerge until the beginning of the nineteenth century. When it did appear, these three problems were its first victims: all were proved impossible. Gauss was the first to suspect this, but could not prove it. The impossibility of trisecting an angle and of doubling a cube was proved in 1837 by the French mathematician Pierre Wantzel (1814–1848), and the impossibility of squaring the circle was proved by the German mathematician Ferdinand von Lindemann (1852–1939) in 1880.

The underlying reason for the impossibility of all three constructions is expressed in the same theorem. This theorem says that a segment of length a can be constructed only if the number a meets two conditions:

(1) a must be the solution of a polynomial equation with integer coefficients.
(2) Among all the polynomial equations for which a is the solution, the equation of the lowest order is of an order that is a power of 2 (that is, the order must be 1, or 2, or 4, or 8, etc.).

Examples:

(1) Take the equation $a^2 - 2 = 0$. This is a second order equation, and its order, 2, is a power of 2. Its solution, which is $a = \sqrt{2}$, is therefore constructible. Indeed, we saw that the square root of any already-constructed length can be obtained with a ruler and compass. In the chapter "Mathematical Harmonies" we also saw a very simple way to construct a segment $\sqrt{2}$ long: it is the diagonal of a square with a side of length 1.
(2) The equation $a^4 - 2 = 0$, too, fulfills the above conditions, since it is of order 4, which is a power of 2. Sure enough, it is possible to construct a segment of the length that is its solution, that is, of length $\sqrt[4]{2}$. The reason: $\sqrt[4]{2}$ is the root of $\sqrt{2}$. The latter we can construct, and hence we can construct its root.

How does this theorem lead to the conclusion that squaring a circle is impossible? Let us take the version of the problem that speaks of circumference: "Construct a segment the length of the circumference of a circle with radius 1." The circumference of a circle with radius 1 is 2π. If we knew how to construct a segment of length 2π, by halving it we would obtain a segment of length π. I already mentioned (in the chapter "The Power of the Oblique,") that in 1768 Lambert proved the irrationality of π. In 1880 Lindemann proved even more: that π is "very irrational," or in technical terms "transcendent" or "non algebraic." This means that it is not a solution of *any* polynomial equation with integer coefficients not all of which are 0, let alone the solution of a polynomial satisfying the conditions above. So, by the theorem, π cannot be constructed.

What about the other version of squaring the circle, that of constructing a square whose area is that of a circle of a given radius? Taking the radius to be 1, the area of such a circle is π, and therefore the length of a side of the square must be $\sqrt{\pi}$. But with a compass and ruler we can construct from a pair of segments a segment that is the length of their product, and so, from a segment of length a we can also construct a segment of length a^2. If we were capable of constructing a segment of length $\sqrt{\pi}$, we would therefore be able to construct a segment of length $\sqrt{\pi}^2$, which is π. But, as we already know, this is impossible, and so a segment of length π cannot be constructed.

As for doubling the volume of a cube, assume that we can double the volume of a cube with a side of length 1, whose volume is $1 \times 1 \times 1 = 1$. Twice 1 is 2, so presumably we would succeed in constructing a cube of volume 2. Let x be the length of this cube's side. The volume of a cube with a side x is x^3, so x would satisfy $x^3 = 2$, or $x^3 - 2 = 0$ (in other words, $x = \sqrt[3]{2}$). The equation $x^3 - 2$ is of order 3, and there is no equation of a lower order that $\sqrt[3]{2}$ solves. Since 3 is not a power of 2, condition (2) then rules out the possibility of constructing a segment of length $\sqrt[3]{2}$.

The last problem was the division of an angle into three equal parts. If this could be done with a compass and ruler, it would be possible, among other angles, to divide an angle of 60° into three equal parts, that is, to construct an angle of 20°. It would then be easy to use these tools to construct a right angled triangle with a hypotenuse of length 1, one of whose angles is 20°. It can be proved that the side of such a triangle solves third order polynomial equations; and 3 is the minimum order of equations that they solve. Since 3 is not a power of 2, according to part (2) of the theorem, such a segment cannot be constructed.

Since the proofs of the impossibility of the three classical construction problems, many other impossibility results have been obtained. These are usually hard and deep theorems. There is a straightaway approach to show that something can be done: just do it. It is more difficult to demonstrate that something cannot be done.

Infinitely Large

The eternal silence of these infinite spaces frightens me.

Blaise Pascal, 1623–1662

The riddle of hyperbole

The artist breathes tranquility even in his anxiety.

Nikolai Gogol, 1809–1852, Russian author

Poetry is the spontaneous overflow of powerful
feelings from emotions recalled in tranquility.

William Wordsworth, English poet, 1770–1850

Poetry makes every effort to be indirect. Like any art, it distances itself from its subject. Its language is generally understated, muted, indirect, symbolic, metaphoric, and pictorial. All these means enable poetry to touch things that direct approach cannot. But this is not always so. At least one poetic device seems to do the exact opposite: it intensifies emotions and feelings. I am speaking of hyperbole. The Greek word means "throwing too far" ("hyper" means "too much," and "vole" means "throw"). Mathematics, too, borrowed this term, to describe a curve that is "thrown" to infinity — a hyperbola is one of the conic sections mentioned in the chapter "The Miracle of Order." In poetic hyperbole, things are taken to an extreme and assume colossal dimensions. Here it is in the poem "Stop All the Clocks" (also known as "Funeral Blues") by W. H. Auden, that became famous through its appearance in the movie *Four Weddings and a Funeral.* Although it was originally written as a parody of a eulogy for a politician, the poem was also meant to be serious — which is the way it is perceived today.

Stop all the clocks, cut off the telephone,
Prevent the dog from barking with a juicy bone,

Silence the pianos and with muffled drum
Bring out the coffin, let the mourners come.

Let aeroplanes circle moaning overhead
Scribbling on the sky the message He is Dead,
Put crepe down round the white necks of the public doves,
Let traffic policemen wear black cotton gloves.

He was my North, my South, my East and West,
My working week and my Sunday rest,
My noon, my midnight, my talk, my song;
I thought that love would last forever; I was wrong.

The stars are not wanted now; put out every one:
Pack up the moon and dismantle the sun;
Pour away the ocean and sweep up the woods:
For nothing now can ever come to any good.

W. H. Auden, "Stop All the Clocks"

Resolving the contradiction

It does seem confusing. What is right: understatement or exaggeration? Distance, or intensified emotions? Can hyperbole be an exception among the poetic devices, the only stratagem that puts the cards on the table, and even more so, takes the message to an extreme? If this were so, we would have to abandon the idea of indirectness as the common denominator of poetic techniques. Fortunately, there is no need for this. Actually, hyperbole is not different in this respect from all other poetic devices. Like the others, it, too, is a tool for detachment. It does so by imparting a larger-than-life dimension to the experience. The proportions that things receive are beyond our usual receptive ability, and are therefore experienced in a detached way. When the poet's mourning is shared by the entire universe, it is no longer his. When things are larger than life, a person doesn't really feel them. The suffering becomes the world's, and not his own.

This is best seen in the everyday use of hyperboles, for they are endlessly used (here is one example: "endlessly") in spoken language, as well. "I'd die for some chocolate," "I could eat a horse," "this headache is killing me," "I'm at my wit's end" ("I'm at my wit's end, the ants have even started climbing up the refrigerator"), "a ton of money." What purpose do these hyperboles serve? There is some truth in an exaggerated metaphor such as "this headache is killing me" — it means that it hurts more than I can bear or, in other words, more than I can handle with my usual tools.

The same holds true for hyperbole in poetry. Take, for example, the poem "See the Sun" by the medieval Hebrew poet Solomon Ibn Gabirol (1021–1058). Ibn Gabirol did not have an easy life. He was orphaned at a very young age, and suffered throughout his short life from a terrible skin disease and torturous digestion problems. He had one consolation in life — the support of a rich benefactor named Yekutiel. "See the Sun" is a lament on the death of his beloved patron.

See the sun gone red toward evening
as though it were wearing a crimson dress,
stripping the edges of north and south
and, in violet, lining the wind from the west:

and the earth — left in its nakedness —
takes refuge in the shadow of night, and rests,
and then the skies go black, as though
covered in sackcloth, for Yequtiel's death.

Solomon Ibn Gabirol, "See the Sun," trans.
P. Cole, *Selected Poems of Solomon Ibn Gabirol*

This amazingly modern poem derives its force from three poetic devices. The first is the twist at its end. The poem's true meaning is revealed only in the last line. One of the later chapters of the book will be devoted to this device. Here let me just explain that the beauty of a twist lies in the fact that everything that came before suddenly receives new meaning, and the reader has to absorb and comprehend a great deal all at once. It is only in the last line that the reader realizes that the portrayal of the sunset and the earth's departure by the sun is only a metaphor for the poet's sense of abandonment upon the death of his patron. We must then decipher anew all the preceding lines. Since conscious thinking cannot absorb so much so quickly, the understanding remains partially subconscious. The second device of the poem is displacement, the offhanded statement of the crux of the matter. Yequtiel's death is mentioned as part of a metaphor: "as if it is covered by sackcloth." Very similar to the displacement in Lea Goldberg's "About Myself," in the chapter on displacement at the beginning of this book. Yet, the strongest device is probably the third one: hyperbole. The pain at Yequtiel's death is attributed to the entire world — to the earth, the skies, the sun. Projecting his mourning to the skies, it is easier for the poet to bear his pain.

The beauty of hyperbole resembles that of a majestic landscape. A soaring cliff, or a tremendous mountain, are not fully comprehended by the viewer. We are accustomed to perceiving the world around us in practical terms, of action, and the cliff or the mountain are too great to even imagine climbing. Similarly, poetic hyperbole transports matters beyond ordinary perception, resulting in absorption without conscious understanding.

Cantor's Story

O, infinity! the most fascinating of mathematical concepts.

David Hilbert

Infinity is the place where things happen that don't.

An interpretation of the concept of infinity,
attributed to an anonymous student

Georg Cantor (1845–1918), German mathematician; the founder of set theory.

A mathematical feud

Mathematics, too, has its hyperboles, things that are bigger than the dimensions of the world to which we are accustomed, and so the rules that prevail in them appear strange and wondrous. This is the concept of infinity. The Greeks already were charmed by this notion and the paradoxes that it creates. But the stormiest turning point occurred at the end of the nineteenth century — stormy in the literal sense, and not only mathematically. The concept of infinity became a battleground.

Except for quarrels regarding claims of priority of discovery, there are hardly any disputes in mathematics. A prominent exception is the story of set theory, a field that was developed in the late nineteenth century by Georg Cantor (1845–1918). Cantor's idea was so novel and surprising that the mathematics community needed about twenty years to digest it. In those two decades wars were waged, polemical articles written, and blood was shed, almost literally. In these wars, several of the period's leading mathematicians took the wrong side.

Cantor was not a quarrelsome person, nor did he intend to overturn the established order. He was engaged in a classic field known as "Fourier analysis," that was originally developed to analyze wave patterns. Cantor made important contributions to this field, but not revolutionary ones. One day, however, for one of his proofs he needed the following fact: that **there is more than one kind of infinity**. There are large infinite sets, and there are even larger. Before Cantor, mathematicians regarded all the infinite sets as equal. All were thought to be very big, and that is all. No one tried to classify infinite sets by size. Cantor showed that, not only was such a classification possible, it was also productive and important. Just as one number can be larger than another, one infinite set can be larger than another. This was the starting point for a new theory — set theory.

The fundamental concepts of set theory are surprisingly simple. Everything evolves from a single concept: an element's belonging to a set. This simplicity might have been responsible for the mathematical community's reluctance to accept the new theory. About fifty years later, John von Neumann would demonstrate that all of mathematics can be formulated within the framework of set theory, but when Cantor introduced his ideas it was hard to believe that such a simple concept could be used to say anything of importance.

Mainly, however, it was another idea that had Cantor's contemporaries up in arms: regarding infinity as a tangible entity. The mathematicians of the nineteenth century were proud of their recent success in providing a rigorous foundation for differential and integral calculus. Since its discovery in the seventeenth century, calculus had proved to be extremely useful, second only to the concept of the number. But for almost two centuries this field rested on shaky foundations, its terms being only vaguely defined. The central concepts in calculus, "tending to a limit" was left fuzzy. Only in the nineteenth century several mathematicians, notably Augustine-Louis Cauchy, Bernhard Riemann, and Karl Weierstrass, provided precise definitions for these terms. In these definitions infinity is something seen from afar but never reached. Tending to infinity means that numbers become as large as we wish, but they never reach infinity. Gauss reproached a friend with whom he corresponded: "I must vigorously protest against the use you make of the term infinity, as something that can be reached. Infinity is only a manner of speaking, meaning numbers that are as large as we wish." To this way of thinking, infinity is "potential," not "actual." Having just exorcised the devil of actual infinity, namely infinite quantities that exist on their own right, the mathematical community was enraged to find Cantor bringing it back through the back door.

Prominent mathematicians, led by the famous Leopold Kronecker, disparaged Cantor's theory, claiming it was worthless. Henri Poincaré, one of the major mathematicians of the time, said that set theory was "a childhood disease, from which mathematics would eventually recover," and accused Cantor of "corrupting the youth" — the same charge for which Socrates had paid with his life some 2,300 years earlier. Unlike Socrates, Cantor was not executed, but he did not receive a coveted university position. The attacks against him further fueled the prolonged depression from which he suffered, and he ended his days in a mental institution. Too late for him to enjoy, but still in his lifetime, his theory was finally victorious, and its importance was universally recognized. Nowadays, set theory is taught in first-year university mathematics courses.

Why numbers are unnecessary to compare sets

The first hurdle that Cantor had to overcome was definition. In order to speak of larger and smaller infinite sets, a precise definition is needed for these terms. But first, an even more basic question must be answered: when are two sets of equal size? For finite sets, the answer is clear: two sets are of equal size if they contain the same number of elements. A set of 5 books is equal in size to one of 5 pencils. But in the infinite case, we don't have numbers at our disposal. Here, Cantor had an extraordinary insight: numbers aren't really needed. Equality of set size can be defined without them, even in the finite case. For example, in order to prove that each of your two hands has the same number of fingers, you don't have to count the fingers on each hand. Just match them, by putting your hands against each other. This is called "correspondence":

> Two sets A and B are of equal size if there exists a correspondence, assigning to each element of A an element of B, in such a way that each element from B corresponds to exactly one element from A.

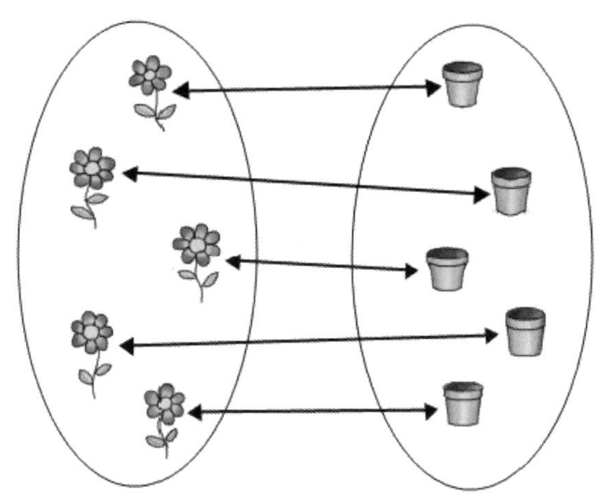

There is a one to one correspondence between the set of flowerpots and the set of flowers, that is, exactly one flower corresponds to each flowerpot, and vice versa. This shows that the two sets are of the same size. This definition skips the concept of number, and can be used for infinite sets, as well.

Another name for correspondence is "function." If the function is named, say, f, then for the element x of set A the corresponding element in B is written as $f(x)$. The correspondence showing equality of sizes of sets is required to fulfill the following: for every element y in the set B there is exactly one element x in the set A for which $f(x) = y$. Just as in the fingers example.

Cantor might have been the first, after thousands of years of mathematics, to realize that numbers are merely intermediaries. The fact that a person has 5 fingers on his right hand means that there is a one to one correspondence between the fingers of his right hand and the set $\{1, 2, 3, 4, 5\}$. This correspondence is effected by counting, that is, passing over the fingers and saying the numbers: 1, 2, 3, 4, 5. A similar correspondence exists between the fingers of the left hand and the same set of numbers $\{1, 2, 3, 4, 5\}$. The two sets of fingers are the same size as that of the numbers from 1 to 5, and therefore are of equal size. Cantor understood that there is no need for mediation: the fingers on the two hands can be matched directly. In the infinite case, when we have no numbers at our disposal, this is the only way to define equality of size between sets. Two infinite sets are defined as being of equal size if there is a one to one correspondence that matches all elements of the first to all elements of the second.

The magic of infinity

A small space has as many parts as a big one.

Blaise Pascal, French mathematician and philosopher, 1623–1662

Cantor's definition of equality of size leads to conclusions that at first encounter seem absurd. Two sets can be of equal size even if one seems patently larger than the other. In the finite case the whole cannot be of equal size to its part. In the infinite case, this is definitely possible, that is, a set may be of equal size to its proper subset. For example, it seems clear that there are more natural numbers than even numbers. Isn't it the case that only half of the natural numbers are even? But this intuition is misleading. In order to show this, have 0 correspond to 0, 1 correspond to 2, 2 to 4, 3 to 6, and so on. Each natural number corresponds to the even number that is its product by 2. Expressed as a function, for every natural number n the corresponding number is $f(n) = 2n$. Note that here we are including 0 among the natural numbers. This is customary, but when convenient we shall omit 0 from the set of natural numbers.

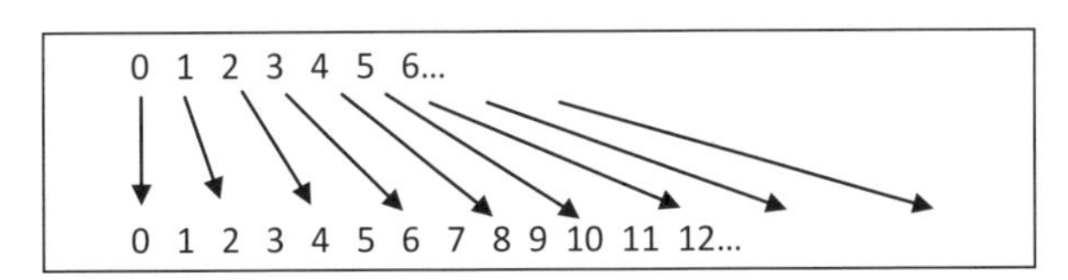

There is a one to one correspondence between the set of natural numbers and its subset, the set of even numbers.

Hilbert put this in the form of a story about a hotel in heaven having infinitely many rooms, numbered as 1, 2, 3, ... One day all the rooms were filled. And then, in the evening, another guest arrived. If the hotel were finite, the guest would have been stranded. But having infinitely many rooms in his hotel, the manager could easily solve the problem. Using a public address system, he asked each guest to move to the next room. The guest in room no. 1 moved to room no. 2, the guest in room no. 2 to room no. 3, and so on. Each guest now has his own room, and room no. 1 is vacant and ready to receive the new arrival.

The hotel is full, but if each guest moves to the next room, a vacancy appears.

The next day, too, the hotel was full. That evening, something even more distressing happened: an infinite number of new guests arrived! But again, the hotel manager did not lose his cool. He asked the guest in room no. 1 to move to room no. 2, the guest in room no. 2 to move to room no. 4, the guest in no. 3, to room no. 6, and on and on. Note that an infinite number of rooms are thereby vacated — all those with odd numbers (1, 3, 5, ...). These rooms can house the infinite number of new guests. For anyone who encounters Hilbert's hotel for the first time, this probably seems mystifying, even entertaining, and possibly even beautiful. I must admit that, even as a professional mathematician, who uses this idea on an everyday basis, Hilbert's hotel still has not lost its charm for me.

And here is another surprise: the set of points on the short segment in the following drawing is equal in size to the set of points on the long segment. One correspondence between them is shown in the drawing. Point Q sends "light beams," and each point on the upper segment corresponds to its shadow on the lower segment.

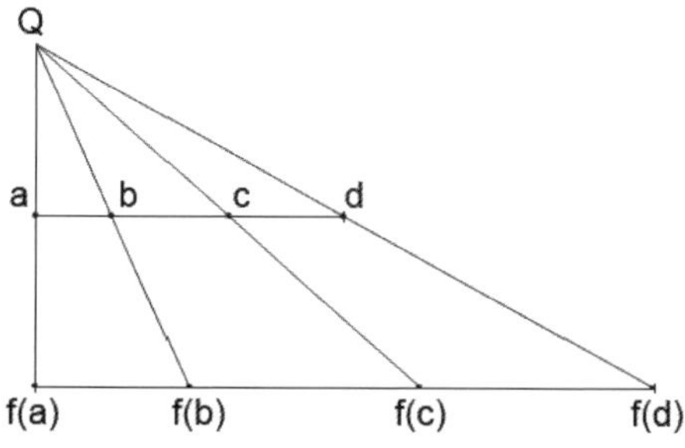

The upper segment has the same number of points as the lower one, despite their being of different lengths. The beams that emanate from a single point establish a one to one correspondence.

Even more surprisingly, the set of points on a finite segment is equal in size to the set of points on the entire infinite line! In order to show this, we will take a segment without its two ends, and bend it into a semicircle. The same idea of light beams creates a correspondence between the points of the bent segment and those of the straight line, as in the following drawing:

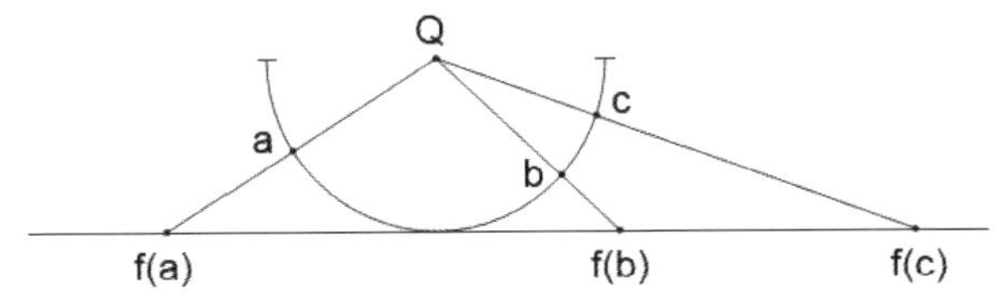

A finite segment with its two endpoints being removed is bent into the shape of a semicircle. The beams issuing from the light source then show that the segment is equal in size to the entire infinite straight line.

Inequality between sets

These examples may create the impression that all infinite sets are of the same size. Cantor's great discovery was that this is not so: there are large sets, and larger ones.

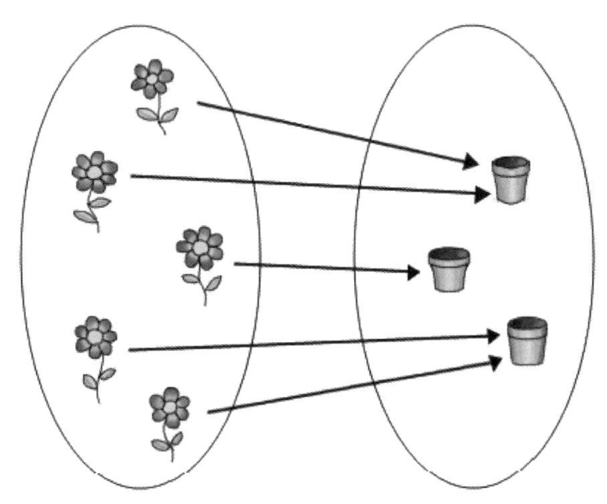

For each flowerpot there is a corresponding flower. This means that the set of flowers is at least as large as the set of flowerpots.

To show this, we have to define inequality between set sizes. Let us begin with the concept "at least as large as ...": when is set A at least as large as set B? Once again, let us begin with the finite case. In the figure above there are 5 flowers and 3 flowerpots. Since 5 is bigger than 3, there are more flowers than flowerpots. This definition, however, is not applicable to the infinite case, because there we cannot count. Accordingly, we will define it, once again, by means of correspondence. The drawing illustrates a correspondence of the set of flowers to that of the flowerpots, that satisfies the following condition: for each flowerpot there is a corresponding flower.

This definition is applicable also for infinite sets: a set A is at least as large as a set B if there is a correspondence assigning to every element of A an element of B, such that all of B is covered. That is, for each element in B there is at least one element in A corresponding to it.

The next drawing, for example, illustrates why the set of points of the segment [0,1) (that is, all the points between 0 and 1, including 0 but not 1) is larger than or equal to the set of natural numbers. For each natural number there is a corresponding point from the segment (actually, many points correspond to each number).

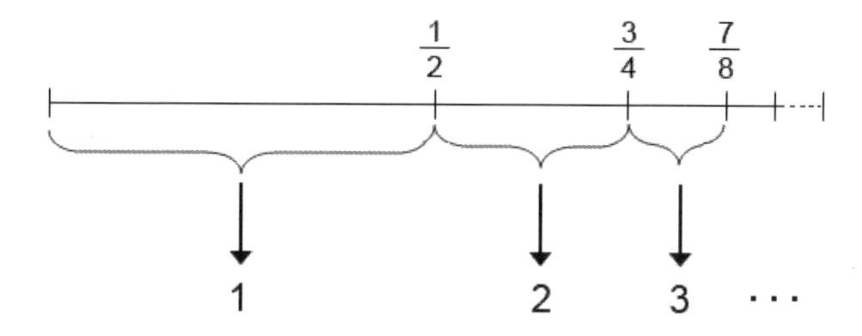

Countable and uncountable sets

A set is called "infinite" if it contains a sequence of distinct elements, say a_1, a_2, a_3, ... In other words, a set is infinite if to each natural number we can assign a different element in the set. If there is such a sequence in set A, then A can use the sequence to cover the natural numbers; that is, A is at least as large as the set of natural numbers. Simply assign 1 to a_1, assign 2 to a_2, assign 3 to a_3, and so on. This covers all the natural numbers, using elements of A, which, by the definition of inequality between set sizes, means that A is larger than or equal to the set of natural numbers.

The conclusion is that every infinite set is greater than or equal (in size) to the set of natural numbers. In other words, the set of natural numbers is the smallest infinite set. An infinite set of exactly the same size as that of the natural numbers is called "countable." Cantor's first great discovery was that there are infinite sets that are not countable. That is, they are actually greater than the set of natural numbers. This discovery is so basic that it deserves a chapter of its own.

The Most Beautiful Proof?

No one shall expel us from the paradise
that Cantor has created for us.

David Hilbert

There are several parameters to the beauty of a mathematical proof. To be beautiful it should be short, surprising, relating to an important and deep fact, and applicable to other problems, from different mathematical fields. Besides the element of surprise, none of these is a necessary condition by itself. Brevity is not essential — a lengthy proof may have the beauty of an edifice. Nor is centrality a cardinal condition — there are beautiful solutions for peripheral problems, too. And finally, beauty is not dependent on utility. But a proof that fulfills all four of these conditions is definitely beautiful.

Only a few proofs meet all these conditions. Cantor's proof is one of them. I suppose that there is not a single mathematician who would not put this proof in one of the first places in his list of the most beautiful proofs. It is short and important, it is used in many fields, and the method on which it is based was the starting point for a whole mathematical field. Despite its simplicity, it can cause shivers in mathematicians who see it for even the hundredth time. Cantor proved the following general theorem:

For every infinite set, there is a larger set.

The diagonal method

Say it all, but say it slant.

Emily Dickinson

Cantor arrived at his theorem in two stages. In the first stage, he proved only that not all infinite sets are of equal size. As already mentioned, the smallest infinite set is the set of natural numbers, and Cantor proved that there is a larger set. In other words, **there are sets that cannot be counted.** "Uncountable," they are called.

Cantor's proof is explicit. Not only does it show that there is an uncountable set, it also expressly presents such a set. This is the set of sequences composed of 0's and 1's. There are many such sequences. For example: 0, 0, 0, 0, ...; 1, 1, 1, 1, ...; 0, 1, 0, 1, 0, 1, ...; or 0, 1, 0, 0, 1, 0, 0, 0, 1, 0, 0, 0, 0, 1, ... Cantor showed that there are not just many sequences like these, there are very many, more than the natural numbers.

The proof is by negation. That is, upon assuming that the theorem is false, a contradiction is reached. Let us assume, Cantor said, that it is possible to count all the sequences whose terms are 0 and 1. The enumeration below is just an attempt, whose purpose is to make things concrete:

$S_1 = 0, 0, 0, 0, 0, 0, 0, \ldots$
$S_2 = 1, 1, 1, 1, 1, 1, 1, \ldots$
$S_3 = 0, 1, 0, 1, 0, 1, 0, \ldots$
$S_4 = 0, 0, 1, 1, 0, 0, 1, \ldots$
$S_5 = 1, 0, 1, 1, 0, 1, 1, \ldots$
$S_6 = 1, 0, 1, 0, 1, 0, 1, \ldots$
$S_7 = 0, 0, 0, 1, 1, 1, 0, \ldots$
....

The assumption is that all 0, 1 sequences should appear in this list. Namely — that we succeeded in counting all of them. But now, Cantor said, I will show you that in actuality you must have failed to count them all. I will show you a sequence that definitely does not appear in this list. To this end, look at the diagonal of the table:

$S_1 = \mathbf{0}, 0, 0, 0, 0, 0, 0, \ldots$
$S_2 = 1, \mathbf{1}, 1, 1, 1, 1, 1, \ldots$
$S_3 = 0, 1, \mathbf{0}, 1, 0, 1, 0, \ldots$
$S_4 = 0, 0, 1, \mathbf{1}, 0, 0, 1, \ldots$
$S_5 = 1, 0, 1, 1, \mathbf{0}, 1, 1, \ldots$
$S_6 = 1, 0, 1, 0, 1, \mathbf{0}, 1, \ldots$
$S_7 = 0, 0, 0, 1, 1, 1, \mathbf{0}, \ldots$
....

Now, write the sequence that appears on the diagonal: $S = 0, 1, 0, 1, 0, 0, 0, \ldots$ and change every 0 in it to 1, and every 1 to 0. This gives us the sequence T = 1, 0, 1, 0, 1, 1, 1, ... What is special about T? It differs from S in each of its terms. Since the first term in S is the first term in S_1, we know that T differs from S_1 in its first term: S_1 has 0 in the first place, while

T is defined as 1. Consequently, T is not equal to the sequence S_1. Similarly, T is different from S_2 in its second place: S_2 has 1 in its second place, and T reversed it. It has 0 in the second place. And if T is different from S_2 in the second place, the two cannot be the same sequence (if two sequences are equal, they are equal in all places). T differs from S_3 in its third place, and therefore is not identical to S_3. And on and on: for each number i, the sequence T is different from the sequence S_i in the ith place. Accordingly, T differs from all the sequences S_i, that is, T does not appear in the list. So, this list does not include all the 0,1 sequences in the world! And this is true for every attempt to count the 0,1 sequences. Every such attempt must fail.

Conclusion: The set of real numbers is not countable

One of the conclusions of this theorem is that there are more real numbers than natural numbers. Every sequence of 0's and 1's can be matched with a real number, by adding a 0 and a decimal point to the left of the sequence. For example, the sequence 0, 1, 0, 1, ... corresponds to the number 0.0101...; the sequence 0, 0, 1, 1, 0, 0, 1, 1, ... corresponds to the real number 0.00110011.... Since the sequences whose terms are 0 and 1 cannot be counted, it is also impossible to count all the real numbers of this type, that is, the numbers in whose decimal representation a 0 appears before the decimal point, and only 0 and 1 after it. These, of course, are only a small portion of all the real numbers, and if even these cannot be counted, then all the real numbers most certainly are uncountable.

For obvious reasons, the idea used by Cantor in his proof is called the "diagonal method." Since its discovery, it has repeatedly proved its effectiveness, and has become a standard mathematical tool. We will return to its later exploits, but first, let us see how it developed into a proof that for every set there is a larger one.

The power set

Two years later, Cantor proved the more general result already mentioned: there is no largest set. Every set has a set larger than itself.

Like the proof for the existence of an uncountable set, the proof is explicit. For every set A, Cantor exhibited a specific set larger than A. This is the **set of subsets** of A. Set S is called a "subset" of A if it contains part of the elements of A. "Part" means any part — S might contain no element at all (that is, it will be empty), or it might contain all the elements of A (that is, it will be equal to A). For a reason that will immediately become clear, the set of the subsets of A is called the "power set" of A. For example, if A is the set of the two terms $\{1, 2\}$, its power set contains 4 sets: $\{1\}, \{2\}, \{1, 2\}$, and the empty set, that contains no element, and that is labeled $\emptyset$.

Cantor proved the following:

The power set of A is larger than A itself.

In the example above, the size of A is 2, while the size of its power set is 4, and so the power set is indeed larger. In a simpler example, if A is the empty set $\emptyset$, then it has a single subset, which is it itself. Its power set is therefore of size 1, while the set itself has only 0 terms. The set $A = \{1, 2, 3\}$ has 3 elements, as compared with its 8 subsets:

$$\emptyset, \{1\}, \{2\}, \{3\}, \{1, 2\}, \{1, 3\}, \{2, 3\}, \{1, 2, 3\}.$$

The source of the name "power set" is evident from these examples. A set of n elements has 2^n subsets, that is, the size of its power set is 2 to the power n. (In these examples: a set of size 2 has $2^2 = 4$ subsets; a set of size 3 has $2^3 = 8$ subsets; and a set of size 0 has $2^0 = 1$ subsets.)

Cantor plays with Mr. Potato Head

How did Cantor prove his theorem? Let me demonstrate, using the familiar toy Mr. Potato Head. This is a toy model to which you can add ears, a mustache, eyebrows.... In our version, each feature has precisely two possibilities — being there or not. So for example, our Mr. Potato Head either has a moustache or not, either has eyebrows or not, and so on. Each choice of features creates a different character: one has a moustache and eyebrows but no ears, and another has ears and eyebrows, but no moustache. So a character is determined by a subset, any subset, of the set of features. Cantor's theorem ("there are more subsets than elements") reads in this case: *There are more characters than features.*

How do we prove this? To show that a set A is larger than a set B we should show that it is impossible to cover A by B. In our case, in which we want to show that there are more characters than features, we should show that there is no assignment of characters to features that covers all characters.

So, suppose that we are given some correspondence, that assigns to every feature a character. I (or rather, Cantor) will show you a character that is not covered, namely it is not assigned to any feature. In fact, I will construct such a character in front of your eyes. Let us call this character Out of Spite — you will soon understand why.

Remember? To each feature there is assigned a character. So, there is a character assigned to the nose. Now, this character may have a nose, and it may not. Out of Spite will do precisely the opposite: if this character has a nose, Out of Spite chooses not to have one. If it doesn't, Out of Spite chooses to have a nose. In any case, Out of Spite is different from the character assigned to the nose. They differ with respect to whether they have a nose or not.

Let us go next to the moustache. There is a character assigned to it. This character may have a moustache, and it may not. If it has, Out of Spite will choose not to have a moustache. If it does not have a moustache, then Out of Spite will choose to have a moustache. Then Out of Spite is different from the character assigned to the moustache — they differ on the moustache.

I think it is clear how Out of Spite is defined, and how he gained his name: for every feature, he chooses oppositely to the choice of the character assigned to that feature. It differs on the hair from the character assigned to the hair, and on the left ear from the character assigned to the left ear. The result? Out of Spite is different from *all* characters assigned to the features. It is not equal to any of them. So, we found a character (Out of Spite) that is not assigned to any feature. Since this argument worked for every assignment, it implies that there is no assignment covering the characters by features. There are too many characters to be covered.

In order to prove Cantor's general theorem, that the power set of every set A is larger than A, just give A as a set of features to Mr. Potato Head. If he is mathematically inclined, he will agree to use any set — say the set of even numbers. We have seen that there are more characters formed by choices of subsets of A than the size of A. That is, the subsets of A are more numerous than A itself. This means that the power set of A is larger than A.

Cantor proved that the number of possible characters of Mr. Potato Head you can form with given features is larger than the number of features, even if the number of features is infinite.

Even after many years of acquaintance with this proof, I am still charmed. The definition of Out of Spite is a rabbit popping out of a hat. The proof is so short and surprising, that it takes some time to digest. It has few competitors for the ratio between importance and length.

Returning to more abstract terms, what Cantor argument says is this. Let A be any set. If f is an assignment of subsets of A to elements, so that to every element a of A we assign the subset $f(a)$, there is a subset C (after "Cantor" — this is the "Out of Spite" set) that is not assigned any element. What is C? An element a belongs to C if it is not included in $f(a)$. This is the "spite." So, C is the set of elements that do not belong to the set assigned to them by f. This is a rather confusing definition. It smells of circularity, which is another name for "self-reference." A bit like defining a man as "the father of the person hereby defined," or a number as "the number hereby defined, plus 1." Such definitions tend to generate paradoxes, like a man who is own father or a number that is greater than itself by 1. And indeed, this is what happened next in set theory. Paradoxes appeared as if from nowhere, threatening to topple the entire beautiful edifice erected by Cantor.

Paradoxes and Oxymorons

A parody of a proof

Cantor's proof fell victim to something rare among mathematical proofs: a parody. And not just any parody, but one that an entire generation of mathematicians treated with all seriousness. In 1903 Gottlob Frege completed the second part of his treatise on the foundations of arithmetic. I already mentioned the sorry fate of the first volume that received scant attention. An even more unpleasant surprise awaited Frege now. He sent the almost finished text to Bertrand Russell in England. Russell sent him a shocking reply: he showed that Frege's assumptions lead to a contradiction, a paradox. Frege attempted over the course of several weeks to resolve the contradiction, and finally gave up. He added a pessimistic appendix to his book, in which he changed the system of proof that he had used, in such a way that it lost much of its power. Frege never wrote the third volume. Russell, who had discovered his paradox some time before receiving Frege's book, continued writing his own book, *Principles of Mathematics*, but he, too, added a similar appendix, with an attempted solution to the problem.

Actually, Russell's paradox was not new. Cantor found it, in a somewhat different formulation, some twenty years earlier. Cantor's formulation went like this: according to the theorem (that we presented in the last chapter), for every set there is a set larger than it. But what happens if we take the set of everything in the world? By the theorem, for this set, too, there is one that is larger than it. But this is absurd. This is the largest possible set — every other set is contained in it!

In this formulation, the contradiction is called "Cantor's Paradox." Cantor himself wasn't very upset by it. He probably understood that it does not pose a significant threat to his theory (as indeed became clear later on). Other mathematicians, as well, did not take it seriously. Russell's paradox, in contrast, did make waves. In those years mathematicians tried to write precise axioms for set theory, and they realized that naïve axiomatization will not do. It leads to contradictions.

Russell's contradiction had greater impact because of its simpler formulation. Cantor relied on a theorem, his theorem that for every set there is a larger one. Russell's paradox is a one line contradiction, constructed from scratch. But as mentioned, the two paradoxes are really the same. Unlike Cantor, Russell was not a creative mathematician, but he had good mathematical education, and he knew what has to be done upon encountering a contradiction: analyze the proof leading to the contradiction. His analysis led to the argument that, actually, is at the heart of Cantor's paradox. Here it is: most of the sets that we can think of do not belong to themselves. For example, the set of chairs does not belong to itself, because it itself is not a chair. The set of natural numbers does not belong to itself, because it itself is not a natural number. But there also are sets that do belong to themselves, for example, the set of all the things in the world, since it is something in the world; or the set of all sets (since it itself is a set); or the set of everything whose name begins with "s."

> Call the set of all the sets that do not belong to themselves R (after Russell). Does R belong to itself? By the definition of R, a set belongs to it if and only if it does not belong to itself. If we apply this rule to R itself, we obtain: R belongs to itself only if it does not belong to itself. This is a contradiction: a proposition and its negation cannot both be true at the same time.

Russell formulated a parable to illustrate the paradox. A barber who lives in a small village vows to give a haircut to precisely those villagers who do not cut their own hair. But now he is in a quandary: must he cut his own hair, or not? If he gives himself a haircut, then, according to the vow he took, he cannot cut his own hair. But, if he will not give himself a haircut, then he must do so!

To cut, or not to cut?

The story of the barber is not a paradox. His vow simply cannot be realized. Russell's set, however, seems to lead to a real contradiction. Is this so? Certainly not. If the mathematical axioms are chosen well, they will not result in a contradiction, because they will describe reality, and in reality there are no contradictions. Like every paradox, this one was born out of a shaky definition. Behind Russell's paradox, as behind the original paradox of Cantor, is an assumption called the "Axiom of comprehension." This states that every property defines a set: the property of "being a chair" defines the set of all chairs; the property "being an even natural number" defines the set of even natural numbers. This assumption, however, enables circular definitions. As Russell's paradox shows, with the help of the axiom of comprehension, we can define a set for which the relationship "belonging to itself" is self-defined.

For a brief moment, the foundations of mathematics shook. It seemed as if the paradox would expel us from Cantor's paradise. But things were soon set right. Ernst Zermelo wrote in 1908 a system of axioms, that was improved in 1922 by Abraham Fraenkel. The Zermelo-Fraenkel system of axioms does not contain the axiom of comprehension. Sets are constructed with greater care, from axioms that (apparently) do not lead to circular definitions. And so, Cantor's paradise remained unscathed, enabling Hilbert to say that the danger of expulsion had passed (his declaration cited above: "No one shall expel us from the paradise that Cantor has created for us," was made in 1926). This is still one of the most beautiful mathematical fields, and the paradoxes no longer trouble it. In fact, the upheaval had a positive side, as well: the paradoxes of set theory led to some of the most fascinating developments in modern mathematics.

Paradoxes in probability

There are lies, damned lies, and statistics.

Mark Twain

A paradox is not a real contradiction. Its trick is concealing a flawed assumption, and drawing from it an absurdity. It is a hidden error, leading to an apparent one. Solving a paradox means exposing the error. Probability is an area in which intuitions are often misleading, and hence it is easy to cheat there and create ostensible contradictions. Of the many paradoxes in probability I chose one, called "the envelopes paradox."

Suppose that your rich uncle presents you with two sealed envelops, and tells you that he put in one of them — you do not know which — a double amount of money as he put in the other. You are free to choose an envelope, which you do. You open it and find a $100 bill. Now your uncle evinces double generosity: "If you want, you can change your choice," he offers. Of course, without first opening the second envelope. The question now is whether it is worthwhile to switch, or not? You randomly chose one of the envelopes, and so there is a $\frac{1}{2}$ probability that you chose the envelope with the smaller sum. In that case, the second envelope contains $200, and switching will give you a profit of $100. There is a probability of $\frac{1}{2}$ that the envelope you chose contains the larger sum, in which case the other envelope will contain a $50 bill, and you will lose $50. So switching the envelopes is a gamble in which there is a 50 percent chance to win a larger sum, and a 50 percent chance to lose a smaller sum — a gamble that is definitely worth taking. And if so, then it is worthwhile to switch.

But this is patently absurd. The argument is not dependent on you having found $100 in the envelope. Had you found $1000, the conclusion should have been the same. But if so, then it would be worthwhile to switch in any case, meaning that you should switch even before you open the envelope. But this, of course, is silly, if only for the reason that it would be worthwhile then to switch again.

The deceit is quite subtle. It is based on an unstated assumption: that each sum has the same probability of being in the envelopes. In our case, in which you opened the envelope and found $100 in it, it is assumed that the likelihood that the uncle put $50 in one envelope and $100 in the other is the same as the second possibility, of $100 and $200. If we were to assume that the probability of the first possibility is much greater than that of the second, you could reasonably expect the envelope you didn't open to contain $50, and so it would not be worthwhile to switch.

The assumption that each sum has the same chance is necessarily erroneous. This is because there are an infinite number of possible sums. When there are (say) 10 possibilities, each with the same probability, the probability of each is $\frac{1}{10}$. When there are infinitely many possibilities, if each had the same degree of probability, then the probability of each of them would have to be 0. But this means that none of them could happen! This means that the probability is not uniform. There are combinations of sums that are more likely than others.

Here is an example clarifying this point. Assume that your uncle does not have more than $2,000,000. In this case, if you find $1,000,000 in the envelope that you opened, you know for a certainty that the second envelope contains half a million dollars, and not two million. In this case, obviously, it is not worthwhile to switch.

Poetical paradoxes

I open the collection of poems *True Love* by the Israeli poet Dalia Ravikovitch (1936–2005), and this is what I find:

Until a little head bursts forth
Red like the orb of the setting sun.

Dalia Ravikovitch, "He Will Surely Come," from *True Love*

The bursting forth is represented by the imagery of its opposite, sunset. Or:

Silence screams within me
And I scream within it.

from "The Beginning of Quiet"

Dalia Ravikovitch, 1936–2005.

Saying something and its opposite in the same breath is called an "oxymoron," that in Greek means "stupid wise man." Samuel Taylor Coleridge wrote in his *Biographia Literaria*:

> [The power of the imagination] reveals itself in the balance or reconciliation of opposite or discordant qualities: of sameness, with difference; of the general with the concrete; the idea with the image; the individual with the representative; the sense of novelty and freshness with old and familiar objects; a more than usual state of emotion with more than usual order [...].

We learn from this that poetry, too, uses paradox. But there is a basic difference: while a mathematical paradox conceals error, there is always truth behind a poetical oxymoron. Beneath the apparent contradiction there is internal logic. By the death bed of the poetess Rachel there were found the lines "Only what I lost/is my possession eternal" (*My Dead*). This contains a truth, because external possessions can be lost. Only what is internalized remains.

Paradox is one of the means poetry uses to maintain tension between surface and interior — the tension that, as I have strived to show, is responsible for the sensation of beauty. Pulling the carpet of logic from under our feet forces us to seek the truth within, and understand that beneath abstract thought there are things no less significant.

Self-Reference and Gödel's Theorem

Hilbert's program

Twenty-four is a good age for making scientific revolutions. Isaac Newton was that age in his miraculous year, 1666, in which he developed the theory of gravity, discovered differential and integral calculus, and formulated the basic laws of modern optics. Einstein was that age when he discovered the special theory of relativity. And so was Kurt Gödel (Austria, 1906–1978) when he proved a theorem that changed our view of mathematics.

The chapter "What Is Mathematics?" described the revolution brought about by Frege, who understood that human thought — and especially mathematical thought — can be studied mathematically. The field that thereby came into existence was called "mathematical logic." Frege's ideas were continued in the early twentieth century by Bertrand Russell, Alfred North Whitehead, and David Hilbert. Hopes soared. Frege taught that it was possible to describe the process of mathematical proof in mechanical terms. He showed that a mathematical proof is just a series of marks on paper that obey well-defined rules. The simplicity of these rules enables us to mechanically examine whether a series of marks is a proof or not. In modern day terminology, a computer could do that. But if this is so, then we can be even more daring: perhaps a computer can also **find** proofs? Can a computer program be written that, upon being given a statement (in itself a collection of marks on paper), will be capable of proving it, or of telling us that it cannot be proved? Just imagine what a wonderful world this would be! Mathematicians could retire, and leave the proofs to computer programs. And even if such a program would not be practical, its mere possibility would be of vast theoretical value.

Questions such as these were asked in the first three decades of the twentieth century. At that time, of course, computers did not yet exist, and in place of "computer programs" people used the term "algorithm," which is a recipe that dictates a precise order of operations, like a recipe for baking a cake. The logicians of that period searched for an algorithm that would check whether a formula is provable or not, and if it decided that there is a proof, it would find one. The main person pushing towards this goal was Hilbert. In fact, he presented his generation with a complete program, consisting of a few challenges. At that time, logicians were mainly interested in number theory. A system of axioms for number theory that was formulated in 1889 by the Italian Giuseppe Peano (1858–1932) was thought at the time to be the last word, and assumed to imply all true statements about numbers. Hilbert's program related to this system. The tasks he set for his fellow mathematicians were:

1. Prove that Peano's system is complete, in the sense that for each formula the system can prove the formula, or its negation.
2. Find an algorithm deciding for every formula in number theory whether it is true in the natural numbers or not.
3. Find an algorithm deciding whether a formula is provable from Peano's axioms.
4. Prove that Peano's axioms are consistent, namely that they do not lead to a contradiction. In fact, the existence of the natural numbers, that obey these axioms, proves this consistency. But Hilbert wanted a proof that relies only on the form of the axioms, not the fact that there is a body obeying them. Such a proof is called "syntactical."

A methodical, but not very efficient, detective

Look again at item 3 above — the search for an algorithm that tests provability. There is a natural candidate for such an algorithm: simply, trying all possibilities. Given a formula you want to check for provability, go systematically, from shorter to longer, over all possible series of symbols. For each such sequence check to see whether it proves the formula, or perhaps proves its negation. Most of the sequences are not proofs at all, but just jumbles of symbols. But perhaps, by chance, like the monkey hitting a typewriter at random, you will hit a proof of the formula or its negation. If there is such a proof, you will get to it at some point. And since checking whether a sequence of symbols is a proof of a given formula is doable, the algorithm is well defined.

Obviously, this is not very efficient. It is like a police detective trying to solve a murder case by examining all the people in the world, one by one. Just as police investigations aren't conducted in this manner, so, too, no one would try to find proofs for mathematical theorems by writing random marks on paper. However, at this stage we are not looking for an efficient algorithm, but for any algorithm at all.

But there is a worse and deeper problem. It is that we don't know when to stop. The algorithm of the murder detective is not efficient, but it is feasible, since there are only a finite number of people in the world, and at some point the algorithm will end. The situation is different for mathematical proofs. If a proof is found at some point during our search, well and good. But if we have examined a million series of signs, and none of them yields the desired proof? Obviously, we could continue on to the one-million-and-first series, but we would never be able to stop and declare: "We exhausted all possibilities, and have not found a proof, so there is none." There is always the possibility that in our next step we would stumble upon the proof. Oil prospectors face this dilemma, but in their case, there is at least a theoretical limit: if they drilled and reached the other side of the earth, this is a clear sign of failure. As far as proofs are concerned, there is no phase in which we should give up.

But note: if we know for sure that one of the possibilities indeed occurs, the formula is provable, or its negation is, then we are in good shape. We can check at each step whether the sequence of symbols at hand is a proof for the formula, or of its negation. Knowing that there exists a proof of one of these, it is guaranteed that at some point our sequence of symbols will be such a proof. So, the algorithm will terminate. So, if item 1 of Hilbert's program is true, namely every formula is provable or its negation is, then we also have an algorithm for deciding which of the two cases it is.

Gödel's theorem and the demise of Hilbert's program

In September 1930 a conference on the foundations of mathematics was held in Königsberg, and was attended by some of the best mathematicians of Europe. An announcement given at the end by a young, shy and slightly-built mathematician went hardly noticed. Luckily, one mathematician did understand the significance of what would be later described as "the most important theorem of the 20th century," and spread the word. It was John Von Neumann, who realized the revolutionary implications of the discovery made by the young mathematician, Kurt Gödel.

Gödel shattered Hilbert's program, in all its details. He proved that the Peano axiom system is not complete; that there is no algorithm for deciding whether a formula on number is true or not; that there is no algorithm for finding proofs in the Peano system; and that there is no syntactical proof of the consistency of the Peano axioms.

The first of these negative results got most of the fame. It is called "Gödel's incompleteness theorem." It is that there are true statements about numbers, which Peano's axioms cannot prove. Of course, since they are true, their negation also cannot be proved. So, there are statements that both they and their negation cannot be proved.

OK, you may say, so Peano was stupid, and didn't devise a good axiom system. Let somebody cleverer come, and propose a better axiom system, that will be complete. Namely, it will decide everything — for each formula, it will prove the formula or its negation. But Gödel's argument has a much wider scope: it is valid not only for Peano's axioms, but also for every reasonable axiom system, where "reasonable" means that it is possible to decide, for every sequence of symbols, if it is an axiom in the system or not.

Gödel's theory drew the attention of a young Englishman named Alan Turing (1912–1954). In addition to his being an exceptionally strong mathematician, Turing also had mechanical skills, and he wanted to give more tangible form to Gödel's arguments. To this end, he invented the first theoretical model of a computer. The actual construction of a computer did not lag far behind. During World War II, Turing participated in building a primitive computer, as part of the effort to break the German code for communications with submarines. Gödel's theory was therefore a significant step toward the creation of the computer.

Kurt Gödel (1906–1978) proved what some consider to be the most important theorem of the twentieth century — that any reasonable set of axioms for number theory is incapable of proving all true facts about the natural numbers.

Circularity

Prayers created God,
God created man,
And man creates prayers
That create God that created man.

Yehuda Amichai, "Gods Come and Go, Prayers Remain Forever," *Open Closed Open*

Gödel was inspired in his proof by a paradox named after the French mathematician Jules Richard (1862–1956). This paradox, like Russell's (in the preceding chapter), is a parody of Cantor's diagonal method. Like Russell's paradox, its deception lies in self-reference. In fact, this is the case in all long-lasting paradoxes. It seems that the human mind is not built to easily detect circularity. The best known paradox based on self-reference is the so-called "Liar's Paradox," invented by Greek philosophers back in the 5th century BC.

This sentence is false.

Think about the truth value of this sentence: if the sentence is true, then, according to its content, it is false. But if it is false, then, again, according to its content, it is true. As in Russell's paradox, we found a statement that is correct only if it is incorrect, which is plainly impossible. Rivers of ink have been poured over this paradox, and philosophers spent many sleepless nights wrestling with it. Actually, the deception behind it is quite simple, and is not very different from the definition of a number as "itself plus 1." The circularly-defined concept in the paradox is the truth value of the sentence. A sentence does not come into the world with a truth value pinned to its collar. In order to calculate the truth value of a sentence, we must do something: compare it with reality. This sentence, however, speaks of its own truth value, and therefore the part of reality to which it must be compared is none other than the truth value itself, that is, the result of the current examination. Thus, the truth value of the liar's sentence is defined by reference to itself. Actually, it is defined simply as its own negation. This is a circular definition, and is therefore invalid. The liar's sentence just has no truth value — it is neither true nor false.

Gödel constructed a paradox of his own, a much more refined one. He considered a statement similar to that of the liar, the difference being that it does not relate to its truth value, but to the possibility of its being proved.

This statement cannot be proved.

Let us name this sentence "G." The following sequence of arguments about G leads to a contradiction:

1. Assume first that G can be proved. Anything that can be proved is patently correct, so in this case G is also correct.
2. But if G is correct, then, according to its content, G cannot be proved, since it states its own unprovability.
3. The previous two arguments show that the assumption that G is provable implies a contradiction. G is then both provable and unprovable.
4. By 3, G cannot be proved, for if it is, a contradiction is obtained.
5. The four first arguments, taken together, show that G cannot be proved.
6. We have shown that G cannot be proved. But by the content of G (that states its own unprovability) we have precisely proved G by this!
7. In 5 we showed that G cannot be proved. In 6 we actually gave a proof for G. These two, together, constitute a contradiction — that is, a paradox.

This paradox is subtler than the Liar's Paradox, and the circularity it conceals is not as simple (a hint: the problem lies in the assumption that "what can be proved is correct," which, if used in proofs, becomes circularly defined). We arrived at a contradiction here, because we looked at the statement formulated in words. Gödel, in contrast, did not arrive at a contradiction. He did not write his statement in words, but as a formula that speaks of numbers — an outstanding achievement by itself. Dressed as a formula, Gödel's sentence does not result in a contradiction, but in a formula that is true in the natural numbers, yet it cannot be proved.

Ars poetic poems

Light didn't just come my way
Nor did I inherit it from my father,
But from my bedrock I bore it,
And hewed it from my heart.

[...]

And under the hammer of my agony
My heart, my rock of strength, shattered
A spark flew to my eye
And from my eye to my verse.

And from my verse it slips forth to your heart,
To fade away in your ignited fire.

And I, with my pith and heart's blood
Pay the price of the flames.

Hayyim Nahman Bialik

Self-reflection appears in poetry, as well. This happens, more than anywhere, in ars poetic poems, in which the poet speaks of his or her poetry. We already encountered a poem of this type, in the passage from "About Myself" by Lea Goldberg. "Light Didn't Just Come My Way" contains two ideas that appear in many ars poetica poems. One is that poetry is not the result of conscious decision, and the poet is merely a passive conduit in the hands of inner forces ("A spark flew to my eye/And from my eye to my verse"). The other is the poet's complaint: while he suffers fiercely, others are entertained by his poetry ("To fade away in your ignited fire").

Here is another poem about the poet's passivity, also by Lea Goldberg, and again, from "About Myself":

Simply:
There was snow in one land
And desert in another
And a star in an airplane window
At night
Above many lands.

They came to me
And commanded me: Sing.
They said: We are words
And I surrendered, and sang them.

Lea Goldberg, "About Myself"

Ars poetica poems occupy a surprisingly large place in poetry as a whole. Should this be attributed to the excessive narcissism of poets? Probably not.

I believe that the answer is not to be found in the personality of poets, but in the beauty inherent in circularity, in having something hang in air because it is hanging on itself.

To end this chapter, here is an amusing example of circularity in poetry. In the poem "A Tale of Two 'Garoos" by the Israeli poet Abraham Shlonsky, the negativist no-'garoo responds to everything with a "No." After learning his lesson, he is asked whether he will remain in his obstinacy.

Will you go on saying "No"?
"No," he answered. "No, no, no!"

Abraham Shlonsky, "A Tale of Two 'Garoos"

Halfway to Infinity: Large Numbers

I will make your descendants as numerous as the stars of heaven and the sands on the seashore.

Genesis 22:17

One of the heroes of the English author David Lodge tries to explain to his friend the meaning of eternity. "Think," he says, "of a ball of steel as large as Earth, and a fly alighting on it once every million years. When the steel ball is rubbed away by the friction, eternity will not even have begun."

Mathematicians wouldn't be impressed by this image. For them, the number of years that will pass until the ball is rubbed away is not especially big. Assume, as an extreme example, that the number of atoms in the ball is that of the number of atoms in the universe, which is estimated at 10^{80}. Assume also that with every alighting, only a single atom adheres to the foot of the fly. After a million times 10^{80}, that is, 10^{86} years, the ball will be rubbed away. Much larger numbers appear in some mathematicians' work every day, especially in my own field, combinatorics. For example, the number of ways in which 100 people can be ordered in a line is much greater than this.

We can deal with such numbers, but they are difficult to comprehend. Even "a million" is a number that people cannot grasp. In the O. J. Simpson murder trial, in which the defendant was charged with murdering his ex-wife and her friend, an expert witness testified for the prosecution that there was one chance in a billion that the blood samples found at the murder scene did not belong to the accused. An expert for the defense claimed that the chance is one in several million, and this statement was enough to acquit Simpson — the jurors didn't have a clue as to the meaning of "one chance in a million."

Mathematicians, too, don't really understand the meaning of large numbers, but they live with them quite well, and they know how to write them concisely. The trick consists of operations that repeat other operations. Multiplication, for example, is a repetition of addition, and raising to a power is a repetition of multiplication. 10^{10} means "10 to the tenth power," that is,

10 times 10 time 10 ... — multiplying 10 by itself 10 times. This is 10 billion, which is written as a 1 followed by 10 zeros. $10^{10^{10}}$ means ten to the (10 to the tenth power), and is written as a 1 followed by 10^{10} zeros. If we were to write this on a strip of paper, the strip would circle the world about 1,000 times.

Is there any meaning to numbers like these? In some conference, a well-known mathematician claimed that there is not. "There is no practical significance to numbers like $10^{10^{10}}$. They are larger than any physical size that will ever be defined. They are also too big to be treated with the regular mathematical tools. We will never be able, for example, to check whether $10^{10^{10}} + 1$ is a prime number."

One member of the audience stood up and asked: "Suppose that two mathematician come to you, one with a proof of Fermat's Conjecture for all the numbers smaller than $10^{10^{10}}$, and the other with a proof of Fermat's Conjecture for all the numbers larger than $10^{10^{10}}$. Which of the results will be more interesting for you?" The lecturer was forced to admit, as would any mathematician, that the second result would be more important. (Fermat's Conjecture was explained above, in the "Unexpected Combinations" chapter. At the time of this story it was not yet solved.)

At this point someone else got up and said: "$10^{10^{10}} + 1$ is not a prime number, because $10^{10^{10}}$ can be written as a fifth power, that is a^5." (For this purpose it suffices for the power 10^{10} to be divisible by 5, which is clearly the case.) The number is therefore of the form $a^5 + 1$, and a number of this form is not a prime. It is the multiple of two smaller numbers, since $a^5 + 1 = (a + 1) \times (a^4 - a^3 + a^2 - a + 1)$ (the reader is invited to check this identity, by opening the brackets). The moral of this story is that, even though we are incapable of grasping the meaning of large numbers, mathematics may still be able to deal with them.

Perhaps because of its name, children are fascinated by the number "googol," which is 10^{100}, written as 1 followed by 100 zeros. Many children think that this is the largest number in the world. A googol is small change compared to the number $10^{10^{10}}$, but still it cannot be comprehended. Once I asked my daughter, "Is there something in the world of which there are a googol?" Without stopping to think, she answered, "Yes. In a second there are a googol googolths of a second" (just as there are ten tenths of a second in a second). Of course, this is cheating — "googolths of a second" exist only in our imagination, we cannot clock them.

There is an additional way in which gigantic numbers appear, presumably in the real world: combinations. For example, a number we already mentioned: the number of ways in which 100 people can be ordered in a line.

How is this number calculated? Each of the 100 people can be put in the first place in the line, and there are then 99 possibilities of putting someone in the second position (only 99, because the first person was already chosen). Accordingly, there are 100×99 possibilities for placing people in the first two positions. For each of these possibilities, there are 98 ways to select the third person, and so, we have $100 \times 99 \times 98$ ways to fill the first three places. Continuing, we find that the number of ways to arrange 100 people is $100 \times 99 \times 98 \times 97 \times \cdots \times 3 \times 2 \times 1$, a number denoted by 100! and is called "100 factorial." Using the formula of James Stirling (in the chapter "A Magic Number"), 100! can be estimated to be about 10^{150}, which is much larger than the number of atoms in the universe. The number of ways in which Earth's inhabitants can be arranged in a line is greater than $10^{10^{10}}$ — so here is this number, in a real-life situation. But, these, too, are not indeed from real life. No one intends to arrange people in all possible ways, nor even to order their names on paper.

Infinitely Small

Everything changes

Everything flows.

Heraclitus, Greek philosopher, ca. 540–480 BCE

Everything flows. The world changes unceasingly — and examining the world means examining its change, like the motion of bodies. Bodies move continuously, and not in jumps — or so it was believed until the appearance of quantum theory about a century ago. And for bodies that are not microscopic, modern physics, as well, assumes continuous change. The mathematical field that examines continuous changes in the world is called "differential calculus."

It is sometimes jokingly said that differential calculus was discovered by someone who believed that the world is flat — and was right. An ant that stands on a large and smooth ball thinks that it is standing on a plane, since from close up the surface appears flat. It was for this reason that people believed that the earth was flat until they possessed the means to distance themselves from it, both in thought and in practice. In a certain sense, differential calculus restores this belief. The assumption on which it is based is that when a smooth line is observed through a microscope, it appears straight. Actually, this is the definition of a "smooth line." When curved lines in the world are examined, they are generally assumed to be smooth. Incidentally, since the discovery of fractals, this assumption is no longer deemed necessary. No matter from what proximity fractal lines are observed, they still appear rough.

Like many fundamental ideas, differential calculus, too, was discovered by the Greeks. They knew how to look at things "through a microscope." The way in which they calculated the area of a circle, for example, looks as if it was taken directly from seventeenth-century differential and integral calculus. They took the circle and divided it into small sectors, as in this drawing:

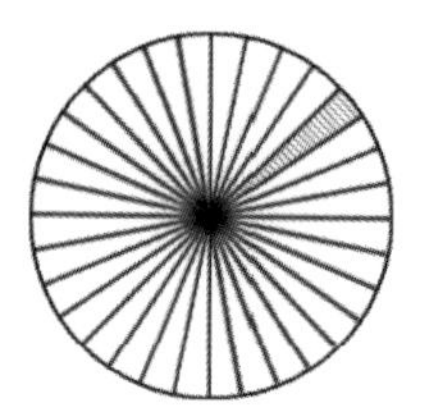

If the sectors are very narrow, to a good approximation they look like triangles. Their base looks like a straight line, just as the earth appears flat to someone observing it from a low height. And we know how to calculate the area of a triangle: this is the base, times the height, divided by 2. If we think of every sector as being almost a triangle, then its height is the radius of the circle. The area of each sector, therefore, is more or less its base times the radius of the circle, divided by 2; and the narrower the sectors, the better this approximation. In consequence, the area of the circle, which is the sum of the sector areas, is — to a good approximation — the radius, times the sum of the base lengths, divided by 2. But the sum of the base lengths is the circumference of the circle. So, the area of the circle is its radius, multiplied by the circumference, and divided by 2. Since the circumference of a circle of radius R is $2\pi R$, its area is $2\pi R \times \frac{R}{2} = \pi R^2$, which is the formula you probably remember from high school. Incidentally, why is the circumference of a circle $2\pi R$? This is a question of definition. The number π is defined as the ratio between the circumference of the circle and its diameter, that is, between the circumference and $2R$.

In antiquity, the use of arguments of this type to calculate areas and volumes was known as the "method of exhaustion." It was developed by Eudoxus, and already appeared in Euclid's *Elements*, that was written in the fourth century BC. The master of this method was Archimedes. He could calculate the area of a circle, the volume of a sphere, and the inscribed area between a parabola and a straight line. Among all his accomplishments, Archimedes most appreciated these calculations. He asked that a cylinder and a ball inscribed within it be engraved on his tombstone, as a testament to the result of which he was most proud, that the volume of the cylinder that inscribes a sphere is $\frac{3}{2}$ times as large as that of the ball.

The idea of "looking through a microscope," that is, using sizes that tend to zero, enjoyed renewed currency in the seventeenth century. This began, actually, by looking through a telescope. Tycho Brahe, the great Danish

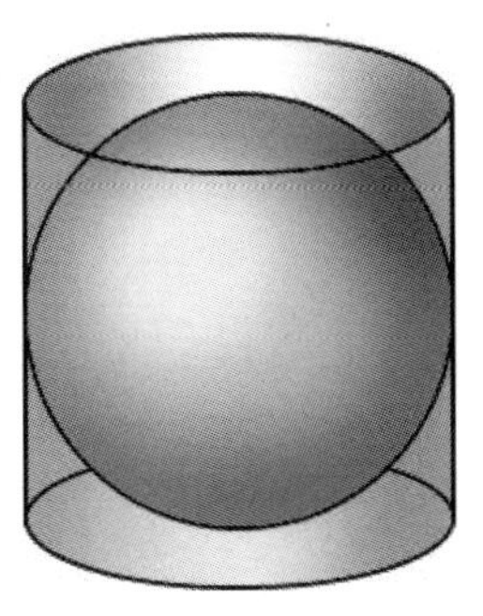

Archimedes asked that this drawing be engraved on his tombstone. He proved that the volume of the ball inscribed in the cylinder is $\frac{2}{3}$ of the volume of the cylinder. He thereby discovered the formula for the volume of a ball.

astronomer (1546–1601), conducted more precise observations than any of his predecessors, and Johannes Kepler, who worked with Brahe in Prague, derived from them three rules that govern the behavior of the planets. The young Newton attempted to explain these rules, and needed a mathematical tool suitable for the study of motion. Modern differential calculus was born out of this need. At the same time, Gottfried Leibniz developed in Germany similar ideas, and also invented the notation that is used in the field to the present day. Newton, who was suspicious by nature, feared that Leibniz was trying to steal his results. He had sent letters to Leibniz in which he presented his ideas, but mail was very slow at that time. Newton believed that Leibniz's slow response was meant to enable him to elaborate Newton's ideas and take credit for them. Newton initiated an inquiry by the British Royal Society, the results of which he himself wrote — judge and jury rolled into one. Notwithstanding this, Newton undoubtedly was the more original of the two. Leibniz himself was exceedingly generous toward his rival, and said that: "Newton is responsible for the better half of mathematics up to his time."

Differential calculus' counterpart is integral calculus. Differential calculus derives information about infinitely small changes from the global behavior of a system. For example, calculating the momentary speed of a body, by knowing its position at any moment. Integral calculus does the opposite: it derives global behavior from the momentary behavior. For example, calculating the position of a body at any given moment, based on knowledge of its speed at any moment.

Archimedes, a Greek who lived in Sicily (287–212 BCE). The greatest mathematician of the ancient world, he was a man of many talents — mathematician, engineer and inventor. He developed a type of pump named after him, invented the planetarium, and built sophisticated weaponry for the defense of his city, Syracuse, against the Romans. When the city was finally conquered, a Roman soldier found him drawing geometric shapes in the sand. As the legend goes, when the soldier asked him what he was doing, he replied, "Don't bother my circles" — an answer that cost him his life.

Representation by a small thing

A poet must have his childhood close at hand.

Theodore Roethke, American poet, 1908–1963

The study of infinitely small sizes has a parallel in poetry, a mechanism called "representation by a small thing." This term was coined by Freud, who discovered it in dreams. In a dream it is used, like all other dreamwork devices, to camouflage forbidden contents. A loaded message is represented by a small and seemingly insignificant detail. This mechanism plays a similar role in poetry: the indirect transmission of the message. The result is poetry's well-known economy of language. Here, for example, is a poem by the Russian poetess Anna Akhmatova (1889–1963). The poetess's confusion is represented by a seemingly inconsequential detail, wearing a glove on the wrong hand:

My breast grew helplessly cold,
But my steps were light,

I pulled the glove from my left hand
Mistakenly onto my right

Anna Akhmatova, "Song of the Last Meeting," trans. in *From the Ends to the Beginning: A Bilingual Anthology of Russian Verse*

Emily Dickinson was a poetess of small things:

To make a prairie it takes a clover and one bee,
one clover, and a bee,
And revery.
The revery alone will do
if bees are few.

Emily Dickinson, "To Make a Prairie" [no. 1755]

Small things, the poem teaches us, leave space for imagination. Since they don't take much space, the rest will be filled by our thoughts. The next poem, by the American poet William Carlos Williams, speaks of the importance of details:

so much depends
upon

a red wheel
barrow

glazed with rain
water

beside the white
chickens

William Carlos Williams, "The Red Wheelbarrow"

This is most likely a childhood picture. Nowhere as in childhood memories is there such a gap between seeming lack of importance and intensity of emotions. A childhood picture, experience, or small object can mean worlds for the grownup. The proportions of a child's experience are different from those perceived by an adult. And, as Theodore Roethke attests, a poet's soul is needed to perceive this in adulthood.

The Japanese, who raised miniaturization to an art form in many fields, did this in poetry, as well. A haiku is a bonsai poem, fashioned of small details loaded with emotion and thoughts:

The dead body
Of a trodden-on crab,
This autumn morning.

Masaoka Shiki, 1867–1902

A small detail can teach of what occurrences outside, and also of what is happening within:

The shadow of the washing-post
Teaches
That winter is in full force.

Masaoka Shiki

Infinitely Many Numbers Having a Finite Sum

The Greek philosopher Zeno of Elea (490–425 BC) was preoccupied with the relation between the freeze frame and the whole picture. Two millennia later, mathematicians would be occupied with the same issues, which would lead them to invent differential calculus. Zeno, who lacked the necessary tools, thought that he had reached contradictions, the most famous of which is "the paradox of Achilles and the Tortoise." The original paradox speaks of a race between Achilles and a tortoise, but I prefer to write a slightly different version, on a competition between the two hands of a clock. The question is this:

The hands of the clock meet at 12 o'clock. At what hour will they meet again?

Zeno's paradox states that this will never happen. This is absurd, but Zeno had a "proof." At exactly 1 o'clock the hour hand will be ahead of the minute one, because it will point to 1, while the minute hand will point to 12. By the time the minute hand reaches 1, the hour hand will advance a bit. Actually, we know exactly how much: $\frac{1}{12}$ of an hour, since its speed is $\frac{1}{12}$ that of the minute hand, and the minute hand advanced from 12 to 1. Now the hour hand points to the hour $1\frac{1}{12}$ (1:05), and the minute hand has to reach this place. But until it does so, the hour hand will advance a bit more. And, once again, we know exactly how much: $\frac{1}{12}$ of the minute hand, that is, $\frac{1}{12} \times \frac{1}{12} = \frac{1}{144}$ of an hour. The minute hand has to reach this place, and in the meantime the hour hand will advance a bit more, $\frac{1}{12} \times \frac{1}{144} = \frac{1}{1728}$ of an hour. This will continue in the same way: every time that the minute hand reaches the previous position of the hour hand, the hour hand will advance a bit more in the meantime. It therefore seems that the minute hand will never be able to catch up to the hour hand! There will always be some gap between them, even if it becomes increasingly small. In the original paradox, in which Achilles ran against a tortoise, Achilles gives the tortoise a head start. According to the exact same argument, Achilles will never be able to catch up to the tortoise, because every time that he reaches the tortoise's former position, the tortoise will have advanced in the meantime, even if only a little bit. But we know, of course, that the minute hand will pass the hour hand, and that Achilles will quickly overtake the tortoise. So where does Zeno cheat?

Before we answer that, let us ask ourselves: at what hour exactly do the two hands meet? We could write an equation, and then solve it. But there is a more elegant, and much simpler, way to do this. Over the course of 12 hours the two hands meet 11 times: every whole hour and something; except for the hour 11 and something, since then the meeting will be at 12, and not at "11 and something." Now, note that the time that passes between each two meetings of the hands is the same. A simple way to see this is to remove the numerals from the clock, and to rotate it so that at the time of a meeting it will look as if both hands are pointing to 12 o'clock. Now it is clear that the exact same time will pass until the next meeting as passed between 12 o'clock and the first succeeding meeting. Therefore, the 12 hours of the day divide into 11 equal parts. And so, $\frac{12}{11}$ of an hour passes between each two meetings. The first meeting after 12 o'clock will be at the hour $\frac{12}{11}$, which is a little before 1:06.

Where did Zeno go wrong? His argument was correct up to a certain point, but he erred in his conclusion. He divided the time to the next meeting of the hands into infinitely many parts. The sum of these periods of time (in hours) is $1 + \frac{1}{12} + \frac{1}{144} + \frac{1}{1728} + \cdots$. This is an infinite series, that is, the sum of infinitely many numbers. Zeno argued that since there are infinitely many terms, the sum is infinite, namely, the time that will pass until the next meeting is infinite. This, however, is wrong. The sum of an infinite number of numbers can be finite, on condition that the numbers decrease at a sufficiently rapid pace. And this is what happens here: each number is smaller than its predecessor by a factor of 12. This means that it is a geometric series, with a quotient of $\frac{1}{12}$ (we first encountered this notion in the chapter "Mathematical Ping-Pong"). And every geometric sequence with a quotient smaller than 1 has a finite sum. The classic example is the geometric series with the quotient of $\frac{1}{2}$, a series in which each element is smaller than its predecessor by a factor of 2. The sum is $1 + \frac{1}{2} + \frac{1}{4} + \frac{1}{8} + \cdots$, which is 2. In order to see this, observe that the distance of 1 from 2 is 1; the distance of $1 + \frac{1}{2}$ from 2 is $\frac{1}{2}$; the distance of $1 + \frac{1}{2} + \frac{1}{4}$ from 2 is $\frac{1}{4}$. The addition of each element in the series halves the distance from 2. Accordingly, the partial sums of the series tend to 2. The "partial sums" are the sums of the first elements — in this case, the first partial sum is 1, the second partial sum is $1 + \frac{1}{2}$, the third partial sum is $1 + \frac{1}{2} + \frac{1}{4}$, and so on.

It is also not difficult to prove that if q is a positive number smaller than 1, the infinite geometric series $1 + q + q^2 + q^3 + \cdots$ converges to a finite number. What number? We can calculate this with the method we used in the chapter "The Book in Heaven." Use the letter S for the sum $1 + q + q^2 + q^3 + \cdots$. Multiply each element in the series by q. This gives us $qS = q + q^2 + q^3 + q^4 + \cdots$. Note how close is the expression for qS to S itself: it just misses the first term, 1. So, $qS = S - 1$. This can be viewed as an equation in the unknown S, that can be easily solved. Moving S to one side we get $S(1 - q) = 1$, and then by dividing both sides by $(1 - q)$, we obtain $S = \frac{1}{1-q}$. For example, if $q = \frac{1}{2}$, for which the series is $1 + \frac{1}{2} + \frac{1}{4} + \frac{1}{8} + \cdots$, we obtain $S = \frac{1}{1-\frac{1}{2}} = 2$, which is what we discovered earlier. In the case of the clock, $q = \frac{1}{12}$, so by this formula the hands will meet after $1 + \frac{1}{12} + \frac{1}{144} + \frac{1}{1728} + \cdots = \frac{1}{1-\frac{1}{12}}$ which is $\frac{12}{11}$ hours, as we already know.

Sequences that tend to 0, with sums that nevertheless are infinite

Is it always the case that when the elements of the series tend to 0, the series sum is finite? The answer is no. The simplest example of this is the following:

$1 + \frac{1}{2} + \frac{1}{2} + \frac{1}{3} + \frac{1}{3} + \frac{1}{3} + \frac{1}{4} + \frac{1}{4} + \frac{1}{4} + \frac{1}{4} \cdots$ (next comes 5 times $\frac{1}{5}$). The elements tend to 0, but two halves are 1, three thirds are 1, and four fourths are 1 — we have a sum of 1 an infinite number of times. This means that the partial sums of the series tend to infinity, that is, the sum is infinite.

The next example is more sophisticated, and also more important, because it appears in numerous contexts. This is the series $1 + \frac{1}{2} + \frac{1}{3} + \frac{1}{4} + \cdots$, which is called the "harmonic series." Its elements tend to 0, but its sum is infinite, which means that its partial sums tend to infinity. In order to see this, partition the sum as follows:

$1 + \frac{1}{2} + \left(\frac{1}{3} + \frac{1}{4}\right) + \left(\frac{1}{5} + \frac{1}{6} + \frac{1}{7} + \frac{1}{8}\right) + \left(\frac{1}{9} + \frac{1}{10} + \cdots + \frac{1}{16}\right) \cdots$. Each pair of parentheses contains a sum that is at least $\frac{1}{2}$. Why? The first pair has two numbers, each of which is at least $\frac{1}{4}$, so the sum is at least $\frac{1}{2}$; the second parenthesis contains 4 numbers, each of which is at least $\frac{1}{8}$, so the sum is at least $\frac{1}{2}$, and so on. So, we have a sum of infinitely many numbers each being at least $\frac{1}{2}$, which gives infinity.

Twists

In jokes, unexpected twists make for a humorous effect. In mathematics and in poetry they generate a sense of beauty. In this chapter I want to return to poems, in which the twist plays a special role: those having a total change of meaning in their last line. We already met such a poem, "See the Sun," in the chapter "Infinitely Large," and discussed the subtlety of the trick.

When the meaning of a whole poem changes in the last line, a lot of information has to be digested in a split second. Suddenly, all previous lines have to be re-interpreted. Since our conscious mind is not quick enough to grasp all these changes so fast, most of the message remains not fully understood.

Let me exemplify this in a poem by the Israeli poet Jacob Steinberg (1887–1947), "The Book of Life."

It may happen that a child, not having a playmate
Would hug a thick book, and though not being able to read
Would leaf through it, page after page.

And then suddenly, as if burdened by some mystery,
  his hands would rest
His tiny fingers clutching to some code, and in his
  eyes there freezes an unanswered hope.
For just a fleeting moment a victorious smile brushes his lips,
And then slowly the tired head falls,
and with a last whining his mouth goes quiet.
Then, just before the child falls asleep, unheeded,
  a hand takes the book away.
An lo, the play is over, like the play of the life of a man.

On its face, this is an uninspired poem. The metaphors are corny, and the story slow and heavy. But then everything changes in the last line. Suddenly we realize that all details of the poem are metaphors for the course of human life. Each line has to be read anew. Life is like a book too hard to understand; man is like a child desperately trying to decipher the code of a book, an attempt that is doomed to fail; fate treats a man towards the end of his life like a forgiving father, that puts his child to sleep at night; the child's sleep turns out to be, in fact, death; the opening words of the poem — "It may happen" acquire an ironical meaning — it is not that "it may happen," it is always the case.

So, the twist is a trick of condensation. A lot of information is compressed into one line. But it is a special type of condensation: it does not require conciseness. On the contrary, the longer the poem is, the more information is compressed into the last line. The details are not understood correctly at first reading, so they may as well be said at length. The moment of illumination will be too brief for a conscious scanning of all these details and reinterpreting them.

Let me just point out one other stratagem that Steinberg's poem uses that makes it a gem: the reversal of roles between tenor and vehicle. Throughout the poem, and in particular in the last line, it seems that life is a metaphor for reading the book. "The play is over, like the life of a man." The meaning is of course the opposite, "life is for us like a book to an ignorant child." "Knowing without knowing," all the way.

Part III: Two Levels of Perception

The most beautiful thing we can experience is the mysterious. It is the source of all art and science.

Albert Einstein

Knowing without Knowing

Men use words only to disguise their thoughts.

Voltaire, 1694–1778, French philosopher

What makes a person beautiful? One factor is essential: that we should not know why he or she is beautiful. Beauty is said to be "blinding," "stunning," "breathtaking" — all expressions attesting to its being beyond our conscious understanding. We may feel stunned by scenery that is too majestic for us to grasp with our ordinary tools of perception. Beautiful musical compositions are too complex for us to know just what happens in them. Beauty is hidden in what we don't completely understand, at least not consciously.

This explains many of the so familiar characteristics of poetry. They are all related to the aim of the poem, of sneaking messages without our notice. The poetic devices are meant to distract our attention, so that the message slips under the radar of consciousness. Brevity, for example, is nothing but the magician's dexterity, meant to deceive our critical faculties. The use of external devices, like rhyme and meter, is meant to draw our attention away from the content. This is the handkerchief under which the magician performs his tricks. Outside appearance hides internal messages: an apparent paradox hides deep truth, while verbally similar phrases may conceal deep underlying contrast.

Poetical repetition

In nonfiction there are few sins more serious than repetition. A piece of information that appears twice in the same text is like a stitch in a garment that is wrongly exposed. "I already know that," the reader thinks, and the magic disappears. In poetry, by contrast, repetition is a powerful device. A famous example is the poem by the Spanish poet Federico Garcia Lorca (1898–1936), "Lament for Ignacio Sanchez Mejias." Mejias, a close friend of the poet, was a bullfighter that was killed in the arena. In every second line the poem repeats the time of the corrida, and the effect is strong — the poem's fame is well-deserved. The following is one stanza from the poem:

At five in the afternoon,
It was exactly five in the afternoon.
A boy brought the white sheet
at five in the afternoon.
A frail of lime ready prepared
at five in the afternoon.
The rest was death, and death alone.

"Lament for Ignacio Sanchez Mejias," *The Selected Poems of Federico Garcia Lorca*, trans. Stephen Spender and J. L. Gili

"Anaphora" is the name given by the Greeks, in the theory of rhetoric, to the repetition of the same combination of words at the beginning of sentences. Epiphora is repetition at the end of sentences, as in this poem. The secret is in the tension between exterior and interior. While on the surface things repeat, underneath there is change. In "Lament for Ignacio Sanchez Mejias," for example, while the words jog in place, the content develops towards a climax, with the tension increasing in each line. The repeated words also convey another meaning: they are like a hammer banging at the mourner's head, forcing him to face the fact he so wishes to deny.

Poetical repetition has an anesthetic aim. When we hear an expression for the second time, we are misled to believe that there is no need to decipher it again. When it comes around a third time, it already all but evades conscious attention. By this the poem attains the magician's goal: diversion of attention. Something very similar occurs in the two well-known poetical repetitions of form, those of sounds (rhyme) and of tempo (meter). The repetitive sounds lull us, and the external similarity between the words leads us to expect like meaning, as well. Thus the poem can sneak in messages below the threshold of consciousness, touching the reader while avoiding drawing his conscious attention to the touch.

Content and Husk

Why poetry?

This poem is a poem of people;
What they think and what they want
And what they think they want.

Nathan Zach, "Intro to a Poem," from *Different Poems*

Poetry is a necessity, not a luxury. It has existed in every society for which there is historical documentation, and it plays a role in everyone's life, whether he or she is aware of this or not. Even the most trite pop song contains a poetical nucleus; our everyday communication is replete with metaphors and symbols; poetry can even be found in road rage curses. But why is this so? What need is there for this strange form of communication, that, on the face of it, does not seem to transmit any information, and frequently is not understood? What place does it fill in our lives?

This chapter will speak of one, well known, answer to this question: poetry opens a window to our inner selves. It enables us to touch deep places within us. This answer is based on a picture of the psyche as consisting of inside and husk. The psyche has a thin shell of logic, that mediates between the inner mental forces and the external reality. This shell is vital for dealing with the world, but it has its price: like every mediator, it creates a partition. It constitutes a barrier that hinders direct access to the inside of the psyche. And denying inner truth naturally creates longings. As a result, man always searches for ways to fool logic, in order to reconnect with his inside. Poetry, and art in general, are one of the means to achieve this. They can dupe the husk, if only for a fleeting moment. Poetry teaches a person that his inner world is no less important than the outer one. It helps him to skip beyond what he thinks he wants to what he wants.

Poetical detachment

As you set out for Ithaka
hope the voyage is a long one

Athika, Constantine Cavafy, trans. Edmund Keeley and Philip Sherrard

In order to penetrate our armor, poetry deludes us. It directs the reader's attention to external elements, in order to facilitate access to inner strata. All poetical devices described in this book are harnessed to this goal. All create a gap between what is on the surface and what is underneath. Seemingly irreconcilable contrasts hide inner logic; combinations that seem impossible, contain truth; metaphors distract attention from the real meaning — all these provide the reader with a life preserver floating on the water for him to grasp when another part of him dives deep down.

The human psyche attaches to the world in many ways, each of which can be detached. When the external hold is severed, room is made for deeper connections. I want to tell here about one type of detachment, with which poetry is especially enamored: detachment of intent or of will. Will is probably man's strongest point of attachment to the world, and therefore its detachment has an especially potent effect. The poem is skeptical of external desires — it attempts to show the heart what are its true longings. The poem by Nathan Zach quoted above (for the second time, but for a different purpose than in the first chapter) is indicative of precisely this detachment. Go slow, these lines say. Stop for a moment — is what you imagine that you want what you really desire, deep down?

A classic example is to be found in ars poetica poems. As I already mentioned, these poems often declare that the poet does not control what he writes. This claim assumes an especially strange form in the poem *After My Death* by Hayyim Nahman Bialik. Not only as regards why he writes poetry, but also as to why he does not, the poet's conscious "I" is not in charge.

After my death mourn me this way:
"There was a man — and see: he is no more;
before his time this man died
and his life's song in mid-bar stopped;
and oh, it is sad! One more song he had
and now the song is gone for good,
gone for good!"

After My Death, Hayyim Nahman Bialik; trans. oldpoetry.com

Bialik, still alive, prophesies that he will never sing his true song, as if this were not at all dependent on his will. His true song, he explains in the continuation of the poem, was destroyed within him, without intent and without his being able to control this.

Poetic justice

History describes what has happened,
poetry says what ***should*** *have happened.*

Aristotle, philosopher, 384–322 BCE

There is wisdom in language. It absorbs and expresses covert processes and hidden thought structures. An expression attesting to the role of detachment of intent in poetry has taken root in all languages: "poetic justice." It was coined by the English critic Thomas Rymer, a contemporary of Shakespeare. Poetic justice is justice appearing from nowhere, recompense that is detached from the sin, but nevertheless suits it. As in poetry, it seems, on the surface, that there is a break between the act and its punishment. Underneath, the truth is revealed: at its best, poetic justice comes from within the person's character.

A recent example that comes to mind is from the Falklands War. In the 1970s a junta of generals seized power in Argentina, and for about a decade they committed atrocious crimes. Thousands disappeared in torture chambers, hundreds were thrown, bound, from airplanes into the sea. No force could withstand the ruling military faction. And then, in 1982, the generals took a foolish step: they took control of the Falkland Islands, windy and insignificant isles in the southern Atlantic Ocean, that were under British sovereignty. Great Britain, headed by Margaret Thatcher, went to war and defeated Argentina. The war was limited to the islands, and did not reach the mainland, thousands of kilometers away. Nonetheless, within a short time after the war the rule of the generals disintegrated in some inexplicable way, to be replaced by a democracy. It might have been the generals' failure in what was supposed to be their strongest side, war, that led to their internal collapse. At any rate, what the popular opposition failed to achieve was done by a senseless war. Justice did not come here from some external authority, nor was it directly connected to the act, but there was truth in it.

Whence the name "poetic justice?" It is called so because something similar to what occurs in a poem takes place within it: the covert, subterranean links are more significant than what is visible. As the poet Percy Bysshe Shelley said, in a passage cited in an earlier chapter, poetry finds the inner similarity between things that, on the face of it, appear different.

Change

Each wave owes the beauty of its line only to the withdrawal of the preceding one.

Andre Gide, author

Connecting with inner forces and deep desires is a well-known aspect of poetry. What is less well-known, and more subversive, is another goal: **change**. One of art's roles is to attain the so desired, and so difficult, aim of changing.

To illustrate how art does this, I will begin with a somewhat prosaic example: putting on shoes. A person trying to put his foot into a stubborn shoe will randomly shake his foot this way and that, without any specific direction. Surprisingly enough, this usually works, and the foot enters the shoe. This is surprising because it isn't clear how chance shaking can be effective. No one mixes random ingredients in a bowl with the hope that the resulting dish will be a success, or just let his legs lead him aimlessly when he wants to get from point A to point B. Why does this work in the case of the shoe? The answer is that shaking up releases from being stuck, that is, from an undesired equilibrium. On the foot's way into the shoe stable situations are liable to be created, that is, situations that are difficult to exit, but nonetheless are not the goal (the foot in the shoe). This is like a person searching for the deepest valley who reaches a shallow one. He might not even know whether he has reached his destination.

If our explorer wants to reach the desired valley, he must be carried away from this temporary point of equilibrium (this is the mathematical term for our local valleys). This is the benefit of random shaking. This idea is used in applied mathematics in a method for finding the minimum of functions. Every once in a while random movements are introduced to escape local valleys.

In life, as in mathematics, sometimes a good shaking is needed to solve a problem. Anyone who has gone through a crisis knows how it can be an impetus for change. Old modes of behavior suddenly turn out to be

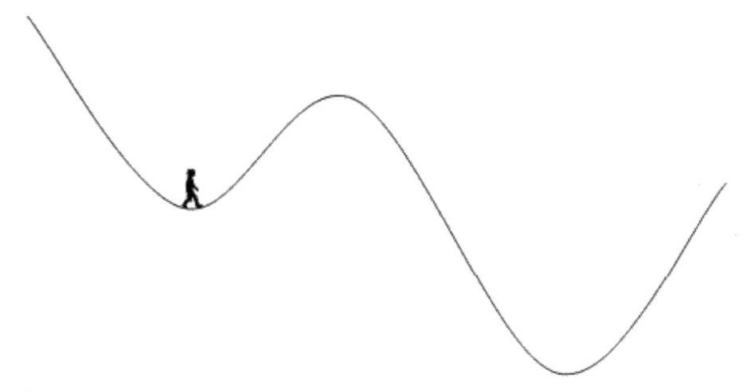

The danger facing someone who is searching for the lowest point: when he reaches a valley, he is liable to be stuck, instead of climbing out of the present valley, to look for the deeper one. In life, too, the conduct that solves a local problem might not be effective in an overall perspective.

ineffective, and a person must retreat within himself to mobilize new forces. Upon returning to the external world, he is likely to use new, often better, behavioral strategies.

But crises exact a stiff price, one that cannot be paid too often. This is why man invented ways to bring about minor shakeups, in which only one attachment to the world becomes undone, gently and not explicitly. The effect is not sweeping, but even a small change is of value. Art is one of the ways to induce such a shaking. The mind sends many minute tentacles to the world, to hold onto ideas, objects, or people. Art removes, if even for a fleeting moment, the grasp of one tentacle. For a moment the energy that was invested in the external world is detached, and retreats inward. When reattached, it can be rebuilt in a new form. A painting that represents its object from a new angle, the reorganization of sounds in music, or the unconventional arrangement of ideas in poetry — all these undo existing attitudes, making us aware of other possible ways of relating to the world.

Detachment and creativity

When asked how he arrived at his discoveries, Einstein replied: "I ignored an axiom." It is not entirely clear to what axiom he referred, because he disregarded many, but most likely he meant the belief that the result of a measurement is independent of the relationship between the measurer and the measured. In the theory of relativity the measured mass of a body, like its length, depend on its speed relative to the measurer. Einstein says that to discover something new, the old has to be relinquished. Sometimes a discovery is not a new idea, but the waiving of an old one. Geniuses like Einstein are capable of giving up old conventions without any external stimulus. Most of us need a shaking that comes from the outside.

Creativity means, before everything, willingness to abandon habitual thought patterns. This is why creativity is so close to humor. Humor is simply the ability to distance ourselves from things. Like humor, creativity demands not taking conventions too seriously.

Know yourself

For childhood
does not grow, no, never.
It is covered with layers, like a thickening shell.

"Noah in the Districts of the Sea," from
Separate Places, Shulamit Hareven

In our psychology-savvy times, it's almost a cliché that change needs self-understanding. In order to change a behavioral pattern, a person must first recognize it. Less well-known is the fact that the opposite is true, as well, and perhaps even more so: in order to know himself, a person must first change. It is hard to be aware of a behavioral pattern if you are immersed in it up to your neck. You first have to give it up, to some degree. The reason for this is that deep personality traits come to the world bound up with their denial. We don't want to know that it is possible to act otherwise. The personality builds barriers against changing.

This means that a shakeup is a tool not only for change, but also for attaining insights. A shock causes a retreat inward, which in turn leads to self-examination. "Unexamined life is not worth living" was Socrates' — perhaps a bit exaggerated — slogan. Since the time of the Greeks, art has been thought to be one of the superhighways for self-observation. Tragedies, for example, give the viewer opportunity to exert his emotions, and thereby come into contact with them, through identification, while not having to experience tragedy in his own life.

Estrangement

Not mine

None of that is mine. I look at it
with surprise. Of whom, from where, all that?

I do not know. An inheritance? No relative or
acquaintance has left me a thing. What now?

Shall I leave this place? If none of that is
mine, perhaps I'll leave this place. And soon?

I don't believe in the question's good faith
and I look at myself with surprise.

"Surprise," from *Other Poems*, Nathan Zach; trans. Yoseph Milman

Science advances by leaps. A scientific revolution takes place every decade or two in almost every field. Anyone who has ever seen the progress of a flock of swallows knows what I'm talking about: the leader of the flock makes a sudden turn, and the entire swarm of birds follows his lead and changes direction. "Quantum leaps" of this sort are made in the natural sciences: a new method or a new concept are invented, and the efforts of the entire scientific community are diverted to the new direction. It seems that progress in the humanities is of a completely different nature. Leaps are rare, and progress is more like the flow of a long and wide river. But there are exceptions, and major discoveries occur also in the humanities. An example in question is a notion discovered in the 1920s by the Russian art scholar Viktor Shklovsky, and named by him "estrangement."

Shklovsky was a theater critic, and a close friend of the poets Anna Akhmatova and Osip Mandelshtam. In some respects he was a lucky person, because, unlike most of those around him, he was not directly affected by the Stalinist terror, and he and his family miraculously escaped the frequent purges. The term "estrangement" that he coined comes from "strange," and means placing things in a strange and new light, for the purpose of restoring pristine freshness to their perception. Shklovsky argued that as the years pass our senses becomes dulled, and they need to be shaken to revitalize them. The role of art, he said, is to cause someone who lives next to the sea to hear anew the murmur of the waves to which he has long been accustomed.

Estrangement is a type of shaking. What is special about it, among other artistic shakeups, is our awareness of it. In most artistic experiences a person forgets himself. Someone watching a movie is usually absorbed in the plot and a person listening to a symphony forgets the world outside. Estrangement does the opposite — it places a mirror before the viewer, and causes him "to look at himself in surprise." The result is the shock of alienation, and the awakening of consciousness to automatic responses. Like the surprise of Monsieur Jordan, the hero of Molière's *The Bourgeois Gentleman*, who suddenly realizes that he has been speaking prose his entire life.

The experience of estrangement is often accompanied by pleasure. For a moment we cease to be slaves to habit, and become its masters. A famous example is the pleasure we derive from discovering the origin of words. Someone who discovers that the three words "radio," "radiator," and "radius" all are derived from the Greek word *radia*, meaning "beam," suddenly becomes master of the words instead of their slave, and entertains the pleasure of freedom. Indeed, one of the terms Shklovsky uses for estrangement is "de-automatization."

Estrangement is particularly characteristic of modern art. Picaso, with his deformed faces, causes us to re-think the way we perceive our surroundings. Stravinsky shocked the ears of the audience of the beginning of the twentieth century, and caused people to stop for a moment and think about what is music and what is the role of sounds in their life. Brecht declared it as an aim, to cause people to step back from the show and realize they are watching theater, and not real life. These are manifest cases, but even when the estrangement is not so obvious it is always there, in all art, if only for the reason that it is taken out of the context of everyday life, and happens in a museum or a concert hall.

Leaving habit behind

Leaving habit behind is essential to the solution of many mathematical problems. Here are some examples, in addition to those we have already met.

1. Is it possible to cut a round cake into 8 pieces with 3 straight-line cuts?
2. Can 4 triangles be formed with 6 matches?
3. The following drawing shows how it is possible to pass through 9 points arranged in a square with a broken line (a line composed of straight sections, that is drawn without lifting the pencil from the page) with 5 sections. Can this also be done with a line of only 4 sections?

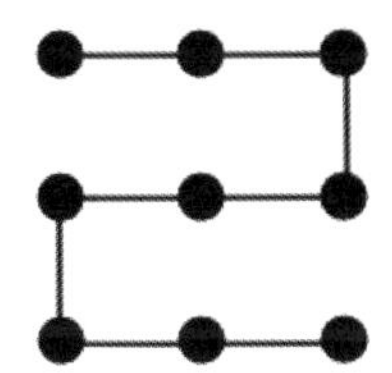

4. A patient has to take daily one pill of type A, and one pill of type B. One day, each of the two bottles had exactly two pills left. And then disaster struck — both bottles fell and broke, and the pills were mixed up. Unfortunately, the two types of pills look the same, and the patient cannot distinguish between them. What will he do?

Habit is the obstacle to solving each of these problems. The first problem is difficult, because we are accustomed to cutting pizzas, and a pizza is divided into 6 sections with 3 cuts, and not 8. But there is a difference: a pizza is thin, while a cake has another dimension in which it can be cut — its height. If we only think in terms of this third dimension, the solution is easy: divide it into 4 pieces by two vertical cuts, as with a pizza; and then cut the cake across its width, perpendicular to the vertical axis.

For the second problem, as well, we have to overcome habit: we are used to two-dimensional matches puzzles, in which the matches lie in the plane. As soon as we forgo this assumption and allow the matches to be in three dimensions, the solution is simple — try it for yourselves.

The difficulty with problem no. 3 is that we assume that the broken line has to be drawn within the square. If we allow it to go beyond the square, then the solution is quite easy. Here it is:

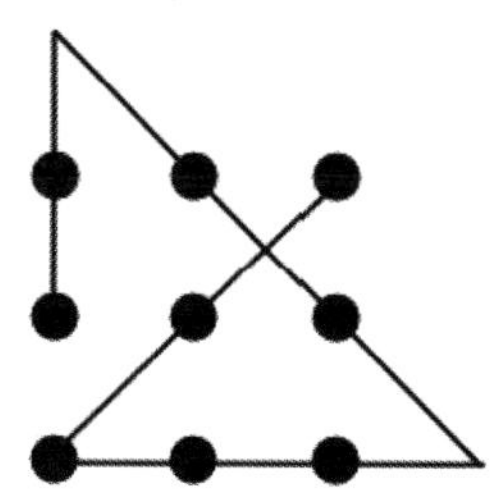

As to the fourth problem, here is a hint: it is not necessary to take whole pills, they can be divided into pieces!

An Endless Encounter

Don't worry about people stealing your ideas. If your ideas are good, you'll have to ram them down people's throats.

Howard Aiken, IBM engineer and inventor

The bridge that is never entirely crossed

Few sensations can be compared to that of making a scientific discovery. All of a sudden, everything falls into place — the examples, the partial results, the hypotheses that were raised on the way. For at least a single moment it seems that the world is going your way. We can only imagine the exultation of Francis Crick and James Watson at the moment of illumination in 1953 when they uncovered the double helix structure of DNA. How, all at once, all facts gathered up to that point coalesced into a coherent whole.

The moment of illumination is supposed to be brief. It is a transition from a state of non comprehension to comprehension, and this can be done only once. But this is not quite the case. As far as beauty goes, this moment lasts forever, as an endless encounter. The joy of discovery remains as time passes, both for the discoverer and for the learner. This is because the transition is never fully consummated. Things are never fully understood. Perception always remains on two levels, along with the wonder at the beauty of the idea.

It will be interesting to compare this with another product of human thought in which surprising shifts play a major role — the joke. There the situation is just the opposite: a joke is like a burnt match. Instead of ongoing wonder and a sense of beauty, it leaves behind a sensation of release, and after being heard the first time, it is no longer funny. This is because in a joke, the bridge is crossed in its entirety. Moreover, after the bridge has been crossed, it is demolished. After hearing the joke, you are on the other side of the river, with no way back. The perspective irrevocably shifts with the punch line, when we realize that the former perspective was merely bait, and is abandoned now.

The turning point in mathematics and in poetry is not a complete change of perspective. We always have one foot on each bank of the river. The reason for this is that a new idea requires its discoverer, and its learner, to establish a complex structure in his brain, and structures are not erected all at once. Usually, they continue for a long time in embryonic form. And so things continue to be a source of amazement even after they seemingly are known. The order in the world is too complex to be understood fully, and the receiver knows that a still deeper order exists beyond the horizon. It is as if what he discovered is only a gold trace on the rock face, that indicates the existence of a richer lode deep in the earth.

Are mathematicians and poets really alike?

On our journey we encountered many points of similarity between mathematics and poetry. But any comparison between the mathematician and the poet fails in one place: their mode of work. Open a window to the mathematician's office and the poet's attic, and look at how they work. For the mathematician, we can see more than one mode of working. Some mathematicians work alone, while others need to talk to someone in order to think better. It also happens that a mathematician's work style changes during his career, usually in the direction of teamwork. As the years pass researchers forge scientific ties, and work more with colleagues or students. Some discover in the course of time that the pleasure derived from working together is comparable to that of a scientific discovery. A common course of events is that at a young age a person engages in research by him (or her)self, making at this stage the most significant discoveries, and as he or she grows older he prefers to work with others. In any case, most of the articles written today in mathematics (and even more so in the natural sciences) have more than a single author, that is, they are the result of joint work. The writing of a poem in company, in contrast, sounds so absurd that if it exists, it is undoubtedly done as a lark.

There is an additional difference between mathematicians and poets. I'll reveal a secret to you, one that is probably hidden from the public, but is well known to families of mathematicians: whether he works by himself or with fellow researchers, most of the time the mathematician stares into space. He tries to draw pictures for himself in his mind, to examine individual cases, formulate hypotheses. He also reads other mathematicians' papers. The poet doesn't engage in any of these. He doesn't read the work of others in order to find inspiration (if he were to do so, this would be transparent). He doesn't invest effort in examining special cases. His staring, so it seems to me, is less intensive, and he works as if in a dream.

All this stems from one basic difference: while the mathematician tries to discover something in the world, the poet's aim is to dive within him (or her)self. The discovery of order in the world can be done in a conversation with someone else, and be aided by ideas of others, but only the individual himself can delve into his soul.

The role of split perception in poetry and in mathematics

There is another difference regarding the two-level perception. In poetry, split perception is absolutely essential. It enables the poet to dive into himself, and bypass the barriers of critical thought. The indirect statement mechanisms are weapons used to penetrate the logical armor that surrounds the psyche like a shell. In other words, in poetry split perception is the aim, and the poetic devices are all means intended to produce it.

It is a different story in mathematics. Split perception is not an aim, but a byproduct. It has no independent role of its own, and is an effect of the discovery process. The sense of magic is only the outcome of the difficulty in consciously digesting the new ideas. As already noted, beauty serves as a motivation for mathematicians. But it isn't the primary goal.

Depth

Maybe its nice, after all, to write poems.
You sit in your room and the walls grow taller.
Color deepens.
A blue kerchief becomes a deep well.

Dalia Ravikovitch, "Surely You Remember," *The Window*, trans. Chana Bloch and Ariel Bloch

But this isn't the whole story. If this were all, the mathematician would not need to have something of the poet in him. There is a more essential similarity between the poet and the mathematician: both look for depth, one in life, and the other in the material world. Both seek subterranean patterns, and to this end both use nonstandard methods of thinking. A new discovery requires momentary disregard of pure logic, plunging into the depths of the unconscious, from where new ideas, both strange and beautiful, are meant to be drawn. In poetry, too, split perception is not just a means, but the result of thought modes of this type. Mathematical magic, like poetic magic, is produced by unexpected leaps of thought, and from amorphous thought in which "a blue kerchief becomes a deep well." This is the matter from which beautiful mathematical discoveries and beautiful poems are cut out. And whoever is capable of such thought could become a perfect mathematician or a perfect poet.

Appendix A: Mathematical Fields

Algebra: Algebra was invented by the Indians, who later handed it down to the Arabs. It reached the West through Al-Khwarizmi's book Al-Jabar, literally meaning "restoration." This refers to a common way of solving equations, by identical operations on the two sides of the equation. The main idea in algebra is calling numbers by names of letters. This is useful in two cases: when we want to speak about general numbers (in this case the letter is called "variable") and when we want to find an unknown number by some indirect information on it (in this role the letter is called "unknown"). The meaning of the word "algebra" changed in the nineteenth century, and came to denote the study of operations, like the four operations in numbers, but more general and abstract. The most basic example of a structure with an operation is the "group", a set with one operation that has similar properties to those of addition in the integers.

Combinatorics, or **Discrete Mathematics:** A field dealing with (usually) finite sets, and relations defined on them. Classical combinatorial problems are about counting: in how many ways can you choose a committee of 4 out of 100 people? "Discrete" is the opposite of continuous, so discrete mathematics usually deals with integer quantities, as opposed to quantities that can be as small as we please. Computers act discretely — a cell can be active or non active, with no intermediate possibilities, and a digit is either 0 or 1, with no third possibility. Hence the study of computer actions needs discrete mathematics. For this reason discrete mathematics has flourished in the last half century.

Differential and Integral calculus: A branch of mathematics that deals with limits, in particular with tending to zero or to infinity. So, in particular it deals with quantities that are "infinitely small" (meaning really — as small as we please), and hence it is also named "infinitesimal calculus." The seeds of this field were sawn already by the ancient Greeks, who used it to calculate areas and volumes. In the seventeenth century the field had a revival, because it is so useful in physics, in particular in the study of motion.

The most prominent among its developers were Fermat, Barrow, Newton and Leibniz. Differential calculus studies continuous change. It goes from the global behavior to the local, for example calculating the velocity of a body at a given moment from knowledge of its overall motion. Integral calculus goes the other way — from local to global. For example, from knowing the velocity of a body at every given moment to the overall distance that it covers in its motion.

For about two hundred years mathematicians studied the notions of tending to zero and to infinity in an intuitive manner. In the nineteenth century it transpired that precise definitions were needed. French and German mathematicians — Cauchy, Riemann, Cantor and Weirstrass, took on this job. They got rid of notions like "infinitesimally small quantities", and replaced them by "quantities as small as we please".

Mathematical logic: This is the "mathematics of mathematics", namely mathematical study of what mathematicians do. The field began with Aristotle, who pointed out some basic rules, and defined what is logical implication. It revived towards the end of the nineteenth century, with the realization of Frege that a mathematical proof is a game that could be played mechanically (in modern day terminology — by a computer). A far reaching revolution in the field was made by Gödel, who proved in 1931 some impossibility results. For example, that no "reasonable" set of axioms for number theory can prove all true statements about numbers, and that though a computer can recognize a proof, no single computer program will be able to prove all provable statements in Number Theory.

Number Theory: The oldest and one of the deepest mathematical field. Its objects, the natural numbers $(1, 2, 3, \dots)$, are seemingly simple, but their study gave rise to the development of entire mathematical fields — for example, algebra.

Set theory: The basic notion of this theory is a very simple one — elements belonging to sets. While the study of finite sets is given to combinatorialists (see above), set theory concerns itself with infinite sets. One of its main topics is the study of sizes of sets. Cantor, the founder of the field, proved that even in the infinite realm there are different possible sizes — there are big infinite sets, and there are bigger.

Topology: Topology is geometry without measuring distances. If you take a rubber sheet, and stretch or compress parts of it in some directions, without making any cuts in it, for the topologist the sheet will remain the same. What matters is only the number of holes in the sheet, which does not change with stretching and compressing.

Appendix B: Sets of Numbers

Natural numbers: The numbers 0, 1, 2, 3, ... These numbers indeed live up to their name. Mathematics takes thought processes and abstracts them, and in this case it is the most basic process that is abstracted: partitioning the world into types of objects. When objects of the same type repeat, we count them "1 apple, 2 apples, 3 apples, ...". The number 0 had a slow start. It reached Europe only in the 12th century.

Integers: These include the natural numbers, plus the negative numbers, $-1, -2, -3, \ldots$. Negative numbers had an even harder time being accepted than 0. They got legitimacy in Europe only in the 16th century.

Rational numbers: Another name for these is simply "fractions". A rational number is a number that can be written as the quotient of two integers, the divisor being non zero. For example, $\frac{7}{3}$, or $\frac{-2}{5}$. The rational numbers arrived much before the negative numbers, because they arise very naturally: even ancient man had to divide apples to 3 equal parts. Pythagoras believed that rational numbers rule the world — every important quantity should be rational.

The reals: Pythagoras was disillusioned. He discovered that not every natural quantity is rational. For example, $\sqrt{2}$ is not. Namely, there is no rational number whose square is 2. So, we have to invent such a number. This number can be approximated by rational numbers as well as we please, meaning that there are rational numbers whose square is as near as we please to 2. In the nineteenth century it was realized also that π, the ratio of the circumference of the circle to its diameter, is also irrational. Of course, this ratio, too, can be approximated by rational numbers. Using approximation, we can construct many irrational numbers. Cantor, towards the end of the nineteenth century, proved that the irrationals are many indeed: they are more numerous than the rational numbers.

Algebraic numbers: These are real numbers that are the solution to a polynomial equation with integer coefficients. For example, $\sqrt{2}$ is algebraic, because it is the solution of the equation $x^2 = 2$.

Complex numbers: The square of a real number is always non-negative. This means that in the real numbers you will not find a number x satisfying the equation $x^2 + 1 = 0$ (in other words $x^2 = -1$). So, such a number must be invented, just like negative numbers were invented to solve equations like $x + 1 = 0$. Such a number, later denoted by i (for "imaginary") was introduced in the sixteenth century. Once i is around, it can be combined with real numbers, namely multiplied by them and added to them. The result of such operations is called a "complex number", for example $3 + 2i$ (complex, because it is composed of two parts, the real part — the 3 in the example above, and the imaginary part — the $2i$ in the example. Actually, we say that the imaginary part is 2, the i being indicated by the word "imaginary"). Gauss showed that using complex numbers, any polynomial equation can be solved. So, there is no need in further extension of the kingdom of numbers, at least not from the direction of seeking solutions for equations.

Appendix C: Poetical Mechanisms Mentioned in the Book

The following definitions of the poetic devices mentioned in the book are titled in the direction of the topic of this book, namely, their effect on the reader and the way they generate beauty.

Anaphora: the repetition of the same combination of words at the beginning of sentences; a special case of the **poetical repetition**, in which an expression recurs throughout the poem. As in many poetical mechanisms, this creates a gap between the external expression and the underlying content: the repetition conceals change. Every time that the poem returns to this expression it has a slightly different meaning, if only for the cumulative effect. Repetition at the end of sentences is called **epiphora**.

Chiasmus: crossing. The word has its source in the Greek letter χ, called "chi" (the parallel of X). Chiasmus exchanges places or roles, such as "He wakes in the morning, but morning doesn't wake in him."

Compression: expressing many ideas with a single symbol. Compression is not a mechanism by itself, and may be effected in many ways, including metaphor, multiple meanings of words, or a picture that contains many messages. This is such an outstanding poetical trait that it is sometimes used as part of the definition of poetry. Poems, like jokes, transmit their messages concisely. The German word for poetry is "Dichtung," which means "compression."

Conceit: a sophisticated metaphor, in which the distance between the tenor and the vehicle (see "Metaphor," below) is large. The word has its source in the originally Latin word "concept." This term was invented to describe the type of metaphors used by the members of a seventeenth-century poetical movement in England, who were called the "Metaphysical poets" by their opponents.

Displacement: turning the spotlight on one element of the poem, while the true message is in an idea that appears at its fringes. That is, displacement emphasizes a less important element in order to incidentally transmit the significant message. Freud discovered this mechanism in dreams, and it also appears in everyday use. For example, magnification words, like "very", are often used for displacement, diverting the stress of the sentence from the main message to the magnifying word.

Hyperbole: exaggeration. The word means in Greek "throwing too far." For example: "I'd die for some chocolate." Like other poetical devices, hyperbole generates detachment. Unlike the other mechanisms, however, that diminish or use indirect statement, hyperbole reaches detachment by distancing the idea. When things are exaggerated and assume superhuman dimensions, we do not sense them directly; it's as if they're from another world.

Metaphor: from the Greek word meaning "to transfer from another place." It uses a usually familiar pattern, called the "vehicle," to describe a (usually) less familiar pattern, called the "tenor." A distinction is sometimes drawn between simile, that uses the words "like" or "as," such as "my beloved is like a young stag" in the Song of Songs, and metaphor, which does not use such words, thereby equating the two terms (the tenor and the vehicle), such as "your eyes are doves" or "all the world is a stage." Actually, the difference between simile and metaphor is minor.

Metaphor is the most widespread poetical device, and is the device most closely identified with poetry. Its power lies in its double role: the transmission of information, and its concealment. It is at the same time an efficient way of conveying ideas, and an indirect way of doing so.

Oxymoron: "stupid wise man" in Greek. The combination of opposites in the same expression, such as "darkness as clear as day," or "silence screams within me." The exterior contradiction generally conceals an inner truth.

Rhyme and meter: like many poetical mechanisms, rhyme and meter turn to the external aspects of words, in order to distract the reader's attention from the inner meaning, thereby enabling it to be delivered subliminally. In this respect, rhyme and meter are close to the mechanism of poetical repetition (see "Anaphora").

Syllepsis: joining together two unrelated ideas, that often belong to different realms. This is a special case of **zeugma**, that combines two things that might or might not be related, such as "I ate the salad and the omelet."

Turnaround: a twist that sheds new light on everything that came before it. The twist usually comes at the end of the poem. It enables compression (see above) without conciseness: at the moment of the turnaround, the reader must absorb, all at once, the meanings of many things that appeared beforehand, and that must suddenly be reinterpreted. Since it is impossible to absorb so much all at once, a large part of the absorption is necessarily subconscious.